4 0000 001 218 462

P9-CEI-173

DISCARDED

WALKING ZERO

By Chet Raymo for Walker & Company

Climbing Brandon: Science and Faith on Ireland's Holy Mountain
The Path: A One-Mile Walk Through the Universe
An Intimate Look at the Night Sky
Skeptics and True Believers

Other Books by Chet Raymo

Natural Prayers
The Dork of Cork
In the Falcon's Claw
Written in Stone (with Maureen E. Raymo)
Honey from Stone
The Soul of the Night
365 Starry Nights
Valentine: A Love Story

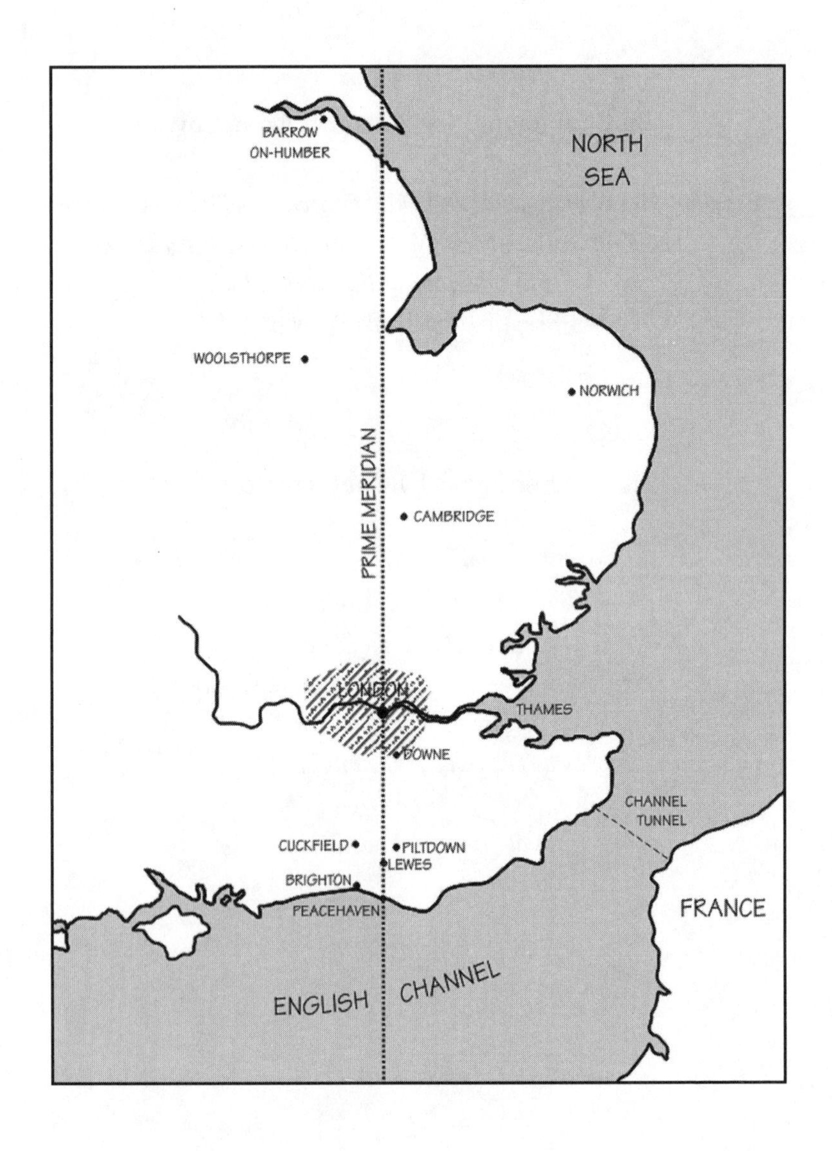
BARROW
ON-HUMBER
NORTH
SEA
WOOLSTHORPE
NORWICH
PRIME MERIDIAN
CAMBRIDGE
LONDON
THAMES
DOWNE
CHANNEL
TUNNEL
CUCKFIELD
PILTDOWN
LEWES
BRIGHTON
PEACEHAVEN
FRANCE
ENGLISH CHANNEL

WALKING ZERO

DISCOVERING COSMIC SPACE AND TIME ALONG THE PRIME MERIDIAN

Chet Raymo

Walker & Company
New York

First published in the United States of America in 2006 by
Walker Publishing Company, Inc.
Distributed to the trade by Holtzbrinck Publishers.

For information about permission to reproduce selections from this book, write to Permissions, Walker & Company, 104 Fifth Avenue, New York, New York 10011.

All papers used by Walker & Company are natural, recyclable products made from wood grown in well-managed forests. The manufacturing processes conform to the environmental regulations of the country of origin.

Library of Congress Cataloging-in-Publication Data has been applied for.

ISBN 0-8027-1494-3

ISBN-13 978-08027-1494-7

Visit Walker & Company's Web site at www.walkerbooks.com

Typeset by Westchester Book Group
Printed in the United States of America by Quebecor World Fairfield

2 4 6 8 10 9 7 5 3 1

There are no privileged locations. If you stay put, your place may become a holy center, not because it gives you special access to the divine, but because in your stillness you hear what might be heard anywhere. All there is to see can be seen from anywhere in the universe, if you know how to look.

—Scott Russell Sanders

CONTENTS

PREFACE

In the fall of 2003 I set out to walk along the prime meridian—the line of zero longitude—across southeastern England. My choice of the Greenwich meridian was not arbitrary; through various accidents of history, the meridian is the touchstone by which the entire world measures place and time. The Royal Observatory at Greenwich, founded by King Charles II in 1675, has defined a common standard for maps and clocks since 1884. The meridian also passes close to a surprising number of other locations important in the history of science. Isaac Newton's chambers at Trinity College, Cambridge, are not far from the line, nor is his place of birth in Lincolnshire. Charles Darwin's house at Downe, in Kent, is two and a half miles from the meridian. And more, much more. It would be hard to think of a walk of equivalent length anywhere in the world that would provide so rich a thread on which to hang a story of human curiosity.

Walking Zero is about the epic struggle to understand cosmic space and time. It is a story of constantly expanding horizons, of intellectual courage and physical adventure, of men and women who dared to believe that the universe was not centered on themselves. It is a story of the breaking of the cosmic egg, of a planet becoming conscious of itself, and of the discovery of an abyss of space and time that might in fact be infinite.

Science is often imagined as a soulless activity administered by men and women in white coats bent on removing spirit and meaning from the world. Nothing could be farther from the truth. Many courageous individuals have bucked reigning orthodoxies to let their imaginations soar where no one had gone before. Pioneers such as Nicolaus Copernicus and Charles Darwin were reluctant revolutionaries who understood that their ideas would be resisted by those who preferred familiar truths. Giordano Bruno, burned at the stake in 1600, paid the ultimate price for (among other things) surmising the multitude of worlds that we now take for granted. Galileo Galilei, as a blind old man, was forced to kneel before assembled churchmen and deny what he knew to be true, the mobility of the Earth.

Our ancestors, perhaps naturally, believed themselves to live at the center of space, coevally with time. "All the world's a stage," wrote Shakespeare, and he meant it quite literally: a stage for the drama of human affairs. Creation myths from around the world assume that the cosmos was created for us, that it is centered on us, and that time has no meaning other than as a frame for human history. Our discovery of *cosmic* space and time—a universe of galaxies and geologic eons that makes no reference to human history—must be counted a triumph of human pluck and cunning. After all, who among us would not *like* to live at the center of the world? Surely, nothing can be more flattering to our sense of importance than to imagine that we are the measure of all things. To forgo the cozy human-centered cosmic egg of our ancestors requires courage and a willingness to contrive our own meaning in a universe that is vast beyond our imagining. In making the journey into cosmic space and time we surrender certainty for curiosity, simplicity for complexity, comfort for adventure. We are perhaps a bit frightened by the light-years and the eons, but we are proud of all that the human mind has come to know and privileged to share, even as spectators, in that epic quest.

PROLOGUE

Each of us is born at the center of the world.

For nine months our physical selves are assembled molecule by molecule, cell by cell, in the dark covert of our mother's womb. A single fertilized egg cell splits into two. Then four. Eight. Sixteen. Thirty-two. Ultimately, 50 trillion cells or so. At first, our future self is a mere blob of protoplasm. But slowly, ever so slowly, the blob begins to differentiate under the direction of genes. A symmetry axis develops. A head, a tail, a spine. At this point, the embryo might be that of a human, or a chicken, or a marmoset. Limbs form. Digits, with tiny translucent nails. Eyes, with papery lids. Ears pressed like flowers against the head. Clearly now a human. A nose, nostrils. Downy hair. Genitals.

As the physical self develops, so too a mental self takes shape, not yet conscious, not yet self-aware, knitted together as webs of neurons in the brain, encapsulating in some respects the evolutionary experience of our species. Instincts impressed by the genes. The instinct to suck, for example. Already, in the womb, the fetus presses its tiny fist against its mouth in anticipation of the moment when the mouth will be offered the mother's breast. The child will not have to be taught to suck. Other inborn behaviors will express themselves later. Laughing. Crying. Striking out in anger. Loving.

What, if anything, goes on in the mind of the developing fetus we may never know. But this much seems certain: To the extent that the emerging self has any awareness of its surroundings, *its world is coterminous with itself.* We are not born with knowledge of the antipodes, the plains of Mars, or the far-flung realm of the galaxies. We are not born with knowledge of Precambrian seas, the supercontinent of Pangaea, or the Age of Dinosaurs. We are born into a world scarcely older than ourselves and scarcely larger than ourselves. And we are at its center.

A human life is a journey into the grandeur of a universe that may contain more galaxies than there are cells in the human body, a universe in which the whole of a human lifetime is but a single tick of the cosmic clock. The journey can be disorienting; our first instincts are toward coziness, comfort, our mother's enclosing arms, her breast. The journey, therefore, requires courage—for each individual and for our species.

Uniquely of all animals, we humans have the capacity to let our minds expand into the space and time of the galaxies. No other creatures can number the cells in their bodies, as we can, or count the stars. No other creatures can imagine the explosive birth of the observable universe 14 billion years ago from an infinitely hot, infinitely small seed of energy. That we choose to make this journey—from the all-sustaining womb into the vertiginous spaces and abyss of time—is the glory of our species and perhaps our most frightening challenge.

Philosophers have endlessly debated the extent to which we can know the world *as it is.* Certainly, we carry an *idea* of the world in our heads, an idea that is partial, tentative, evolving. It is inevitably shaped in its particulars by instinct, experience, culture, and the limitations of human perception. It is hedged around by the bounded capacity of the human brain, which, although wonderfully commodious, is finite. For science to be possible, we must make two assumptions: (1) That the world exists *independently* of our knowledge of it; and (2) that we can know the world with ever-increasing verisimilitude. As obviously true as these statements might seem, in fact their veracity has been long and vigorously debated by philosophers. Nevertheless, they are the foundation upon which all of science rests. The manifest success of science speaks powerfully of the *practical utility* of these two assumptions, if not of their truth.

Science is a *collective* path toward knowledge, a path which—as much as we can make it so—is independent of local cultures, the beliefs of parents and teachers, religion, politics. It is a path that holds the image of the world we carry in our heads against the refining fire of experience. Not just any experience but a special kind of experience called experiment, which, if properly performed and communicated, can be repeated with the same result by any other person equipped with the requisite tools. Science tries as hard to prove an idea wrong as to prove it right. Science requires us to assert our beliefs cautiously, skeptically, tentatively, and be willing to surrender a belief when the collective engine of affirmation fails. Although no one would claim that science is an infallible guarantor of truth, *it is the most effective way the human species has yet devised for making reliable mental images of the world.* As I write, two vehicles, *Spirit* and *Opportunity,* have begun to explore the surface of Mars after journeys across space that required many months. Their journeys would have been impossible had not generations of scientific explorers provided us with reliable knowledge of the world.

I will make one further claim, which is not so much philosophical as personal: The journey toward reliable knowledge of the world *is worth making.* Few people will take exception to this statement. Yet we are adept at finding ways to ignore the world as it presents itself to inquiring minds—the world of the galaxies and geologic time revealed by the long scientific quest. The womb always beckons, with its comfort and security, its total enclosure by a loving, attentive parent. Perhaps we are genetically *predisposed* to favor ideas of the world in which we ourselves are central. None of us, scientists included, is immune to self-deception.

If I am to be your guide for a walk along the prime meridian (and through centuries of the scientific quest), it is perhaps only fair that I reveal something of my personal journey from the womb to cosmic space and time. Like many people, I was raised in a religious tradition still very much grounded in medieval cosmology. The human-centered world described by Dante in his thirteenth-century poem *The Divine Comedy* was more familiar to me as a child than the world of twentieth-century astronomers and geologists.

I was born in Chattanooga, Tennessee, in 1936, of white, middle-class, Roman Catholic parents. Like everyone else, I came into the world unburdened with any but universal human characteristics—certain instincts for survival and inborn behaviors and emotions—but soon I became laden with an increasing burden of cultural baggage. I was southern and therefore predisposed by my environment to racism. Images of burning crosses and men in white sheets were a regular part of my youth, and with those images came an implicit, and often explicit, assertion of the moral and physical primacy of the white race. In church and parochial school I was taught that Roman Catholicism is the one true faith which exclusively holds the keys to salvation. Growing up in the heartland of the Tennessee Valley Authority and New Deal largesse, it was perhaps inevitable that I would later register as a Democrat. The United States of America was, by all accounts, the nation favored by God and by history to be a shining example to the less fortunate peoples of the world. In other words, by accident of birth I found myself situated at the center of a certain conceptual universe, a circumstance I share with every other human who has ever lived.

The subsequent story of my life has been one of moving away from the omphalos—world center—of my birth into a universe of uncentered dimension. For this, I credit mostly my parents, who valued books and history. No one moves from the cozy security of the omphalos untutored and alone. In my parents' library I read of men and women of past and present who dared to peck at the shell of the cosmic egg, who stood on tiptoes to look over the horizon, and who otherwise searched into infinities of space and time. From these pioneers I learned that *my* place, *my* time, *my* race, *my* religion, *my* country were not privileged by centrality, as I had imagined. It was at first a scary feeling—this moving away from the center—akin, I suppose, to what an infant feels upon being separated from its mother. This separation from the center is not something any one of us is likely do on our own without inspiration from those who have gone before.

My own journey began at a relatively young age, even before I was able to read beyond my grade-school primer. Among the books in my parents' library that I particularly remember were a collection of John James Audubon's bird paintings that carried my imagination into the American frontier of the early nineteenth century; a two-volume collection of Currier and Ives prints of tidy village greens that may have been instrumental in my ending up in New England many years later; and a book about Haiti with some spooky and sexually provocative Haitian woodcuts that made me realize the world was bigger and more diverse than what was evidenced by the white-bread, American, middle-class adventures of Dick and Jane. Later, as a teenager, I worked as a shelver in the Chattanooga Public Library and was exposed to further revelations (oh, the glorious *weightiness* of books!) of worlds unseen. Then by a stroke of good luck I won a scholarship to a national university *up north.* Through all of this, books and good teachers expanded my horizons. After earning a Ph.D. in physics I launched myself on a career as a science teacher, filled with idealism—but I was not yet a citizen of the universe.

At that time I was very much in the thrall of the historian of religion Mircea Eliade. In opposition to the quantitative, centerless space and time of Galileo and Newton that I was teaching my students, Eliade offered a cosmos made sacred by hoary tradition. For religious man, space is *fundamentally* centered, said Eliade. In former times, each village, or central hut within the village, had a pole that was presumed to be the axis of the world. Sacred space was *defined* by this local center. Priests and shamans moderated the flow of communication between gods and humans along this central axis. Signs from beyond—hierophanies, or moments of sacred insight—introduced into profane space and time "an absolute element" and put an end "to relativity and confusion." This fixing of a sacred center set up an opposition between *our* world—a centered, orderly cosmos—and the world away from the center—a foreign chaotic place, populated by ghosts, demons, foreigners, and the unsaved. *Our* place, then, is the fixed center of the cosmos, be it the central hut of a village, Jerusalem, Mecca, the bo tree of the Buddha, or the Shining City on the Hill.

For religious man, said Eliade, time, like space, is not homogeneous. It is given shape by a central revelation: God's communication to Abraham, Moses, and the other prophets; Christ's incarnation and death; Constantine's fiery cross in the sky; Joseph Smith finding the golden plates of Mormon on a hillside in New York; and so on. Religious rites and rituals make these singular moments of time infinitely recoverable, infinitely repeatable. For religious man, time is eternally renewed, cycling like the peregrinations of the Sun. Time is reborn, continuously, just as space remains centered, and those of us who reside at the center of the world are reborn again and again with the turning wheel.

Religious time, wrote Eliade, is a "succession of eternities," repeating *again and again* the creation, redemption, and consummation of the world. Likewise, the location of the altar at the transept of the church echoes the location of Eden at the nexus of Earth's cardinal points: north, south, east, and west. (European maps of the Middle Ages placed Eden, or Jerusalem, at the center of the world.) The farther one moves away from the center, the more likely one is to encounter chaos. (At the borders of medieval maps are depictions of monstrous creatures of land and sea.) Sacred time, sacred space—these are ways of maintaining on a cosmic scale the security of the womb in which our bodies and protominds are assembled from the inanimate stuff of creation.

It is perhaps not surprising that I found Eliade's ideas so appealing. I had grown up in a religious tradition in which the ancient cosmic cycles were implicitly celebrated in the daily and annual liturgies of the church. The liturgical day was divided into canonical hours—matins, lauds, prime, terce, sext, nones, vespers, compline—and these segments of the day were sanctified by prescribed prayers from scriptures, thereby weaving the apparently linear thread of Judeo-Christian history into the ever-repeating coil of cosmic time. Even more prominent in Roman Catholic tradition was the colorful wheel of the liturgical year, forever turning, marking out the Sun's eternal return with rites that made no explicit reference to celestial bodies.

And so, following Eliade, I spread a canopy of semipagan religiosity over my scientific studies. From the Protestant theologian Rudolph Otto I absorbed a notion of God as *mysterium tremendum et fascinans* (dread and alluring mystery). I also devoured the works of Joseph Campbell, James Frazer, Emile Durkheim, Lucien Levy-Bruhl, Carl Jung, Bronislaw Malinowski, and other historians and scholars of comparative religion with the same fervor that I read Galileo, Newton, Maxwell, and Schrödinger. There was, obviously, a huge incompatibility in my reading, which because of its inherent contradictions could not sustain itself. The former authors—the scholars of religion—took me round and round, eternally circling the omphalos, the navel of the world, recapitulating the cycles of the Sun and reenacting in ever more modern guise the Myth of the Eternal Return, and the latter authors—the scientists—blazed a linear path that led from the big bang to (perhaps) a final freeze, through an evolving, always-changing, uncentered world in which even human history is represented by a unidirectional arrow of change. I was caught in a bind between enchantment and disenchantment, and the tension was palpable.

If I had to choose a moment when the concept of *uncenteredness* truly dawned on me, it would be in the fall of the academic year 1968–69, when I was supported by a National Science Foundation fellowship to study the history and philosophy of science at Imperial College in London. It was the first time I had been out of the United States and in a culture significantly (although not radically) different from my own. With my wife and three young children, I took up residence in a mews flat just off Exhibition Road in South Kensington, only a short walk from where I would be studying at Imperial College, and even closer to London's Natural History Museum, Science Museum, and Geology Museum. These extraordinary institutions are jam-packed with artifacts associated with the history of science and technology. (I will be visiting certain parts of the collections in the pages that follow.) In the fall of 1968 it was a particular object on an upper floor of the Science Museum that sparked for me a kind of epiphany: the silvered, seventy-two-inch-diameter metallic mirror that for the second half of the nineteenth century was the central component of what was then the largest telescope in the world.

The mirror in the Science Museum had been the light-gathering element of the huge telescope constructed by Lord Rosse, William Parsons, on his estate at Birr, in Ireland, and with which he discovered—among other things—the spiral galaxies. As I recall, this important historical artifact was then displayed by the museum in a horizontal position, like a magical pool of shining quicksilver into which one might look to find the secrets of the universe, a conceit that may have been reinforced by the magical mirrors and pools of childhood fairy tales. And in a sense Lord Rosse's telescope mirror was exactly that. I had previously seen illustrated in astronomy books his famous sketch of the Whirlpool Galaxy in the constellation Canes Venatici, and I had studied enough astronomy by that time to know that the Whirlpool Galaxy is just one of an uncountable number of spiral galaxies, including our own Milky Way, which populate the universe. In my mind's eye I saw the universe of the galaxies reflected there in the glistening eye of the Birr mirror, and I knew that this was knowledge I could not live without.

It would be hard to overemphasize the sudden feeling of *letting go,* the sensation of falling into vertiginous spaces. I knew that the Whirlpool Galaxy is 15 million light-years away. The light of that galaxy's hundreds of billions of stars had been traveling across space—in every direction, not just toward me alone!—for 15 million years before an infinitesimal fraction of the galaxy's photons fell upon Rosse's seventy-two-inch mirror and was brought to a focus by the concave figure of the mirror's surface to form an image on the retina of the astronomer's eye. Here was a line as straight as the arrow of time reaching across the universe, a line that made the cycles of the Earth and Sun—or of a human life—seem paltry indeed. When exposed to the clear night sky during the latter half of the nineteenth century, the glistening face of Rosse's mirror was filled—filled!—with galaxies, nebulas, stars, planets: worlds and worlds without end. There was something in the surface of that metal disk that was indeed *mysterium tremendum et fascinans,* but in a different way than imagined by Otto and Eliade, something that would not be found by clinging to the omphalos of my birth. I knew then that it was not on the axis of the Sun's Eternal Return that I would choose to live my life, but in the quicksilver pool of the telescope, whose depths were capacious enough to embrace an uncountable multitude of worlds.

Finding our way in cosmic space and time is a journey each of us must make alone, and your journey will almost certainly be different from mine, but we build on the achievements of the courageous men and women who have gone before us. Consider for a moment Mary Anning, whom we will meet more intimately in a subsequent chapter. Clad, as her Victorian culture required, in voluminous skirts and bonnet, Anning spent her life digging fossils from the clayey cliffs of Lyme Regis on England's southern shore, not so very far from where I begin my walk. Her efforts revealed the existence of long-vanished seas swarming with dragonlike creatures unlike any animals that exist today. You or I might be lucky in the course of a lifetime to stumble upon a single fossil of one sort or another, and having done so we might wonder what the fossil represents and how it had its origin. Few of us can repeat Anning's achievement—her abundant collection of extinct creatures—because we lack her drive, her talent, her lifelong dedication. Nor can we repeat on our own the accomplishments of the Victorian dons and savants who wrestled with the meaning of Anning's fossilized bones. It behooves us, therefore, to make ourselves aware of the travails and triumphs of our ancestors.

And so, let us begin a journey made partly on the ground of southeastern England with a knapsack and sturdy shoes, and partly in the imagination. Our trek will take us through unspoiled countryside and charming villages, along busy London streets and lazy rural waterways. Along the way we will find the traces of brilliant thinkers and perspicuous observers who over the course of thousands of years added immeasurably to our knowledge of space and time—and opened our hearts and minds to the universe of the galaxies.

1

MAPPING THE EARTH

In 1783 the Prince of Wales, son of King George III of England, visited the seaside town of Brighton, forty-five miles south of London. He was by all reports a wild young man, fond of drinking, womanizing, and gambling. He was sufficiently taken by Brighton's bracing sea air and saltwater bathing to build himself a modest seaside palace.

Over the next several decades his humble "pavilion" was enlarged to become the Royal Pavilion, a goofy and glorious pastiche of onion domes and minarets, in a style some call Indian, but which puts on airs of China, Russia, Arabia, and who knows where else. Needless to say, having a royal prince in town, later a king, gave the town of Brighton a certain panache, and soon a substantial city grew up around the prince's pleasure domes. The railroad came to town, an aquarium was built, and the famous Brighton Pier was extended into the sea. The royal family has long since departed Brighton, but the amenities remain and the city is a favorite destination for Londoners seeking escape from the metropolis.

It is not the pier, aquarium, or Royal Pavilion that brings me to Brighton. The object of my journey is about five miles east of town in the seaside suburb of Peacehaven. The walk there is spectacular—along vertical cliffs of pure white chalk. I have a choice; I can take the path along the cliff tops or the concrete promenade at the base of the cliffs. The promenade was not made expressly for walking; its purpose is to protect the easily eroded chalk cliffs from the force of the sea. Until nineteenth-century engineers built this bulwark, Brighton's history was mostly a matter of watching more and more of the town and surrounding countryside fall into the churning water of the English Channel during winter storms.

I choose the cliff-top walk, and I see what I am looking for well before I reach it: a tall white monument, topped with a globe of the Earth, standing on one of the highest sections of the cliff. When I reach the monument I am at latitude 50° 47' north of the equator and at longitude 0° 0'. *Zero longitude exactly.* I am astride the Greenwich meridian. The inscription of the monument reads:

PEACEHAVEN
KING GEORGE V MEMORIAL
ERECTED BY THE INHABITANTS IN THE
YEAR 1936 TO COMMEMORATE THE
BENEFICENT AND ILLUSTRIOUS REIGN
OF THEIR BELOVED-SOVEREIGN (1910–1936)
AND TO MARK PEACEHAVEN'S POSITION ON
THE PRIME MERIDIAN OF GREENWICH

A smaller plaque, affixed later, bears this inscription:

IN CELEBRATION OF THE
INTERNATIONAL PRIME MERIDIAN CENTENARY
1884–1984
THIS PLAQUE WAS UNVEILED BY THE MAYOR
OF PEACEHAVEN

COUNCILOR ALFA CLAYTON
26 JUNE 1984

What happened in 1884 was nothing less than the globalization of space and time—the fixing of an internationally agreed upon meridian of zero longitude and standard time. Until then, major nations of the world measured longitude with respect to a national capital—so many degrees east or west of London, Paris, Berlin, or Washington. Each country, and sometimes each community within a country, set its clocks at noon when the Sun stood highest in the local sky as indicated by a sundial. There was no uniformity of maps or clocks.

But big things were happening in 1884. Railroads, telegraphs, and empire building were making nations and peoples more interdependent. Messages could be flashed in minutes from Europe to America by undersea cable. Ocean crossings by steamship took days, not the weeks formerly required by sailing vessels. Iron railroad tracks spanned continents. Then, as now, technology was a driving engine of globalization. The pressure became irresistible among nations to arrive at a system of standard longitude and standard time.

On latitude, everyone agreed. There can be no ambiguity about one's location on Earth with respect to north and south. The Earth's rotation defines poles and an equator. If you are standing at the north pole, for example, the stars circle directly overhead as the Earth turns, and the North Star, Polaris, stands near the zenith and barely moves at all. If you are on the equator, the stars arc from east to west across the sky, and Polaris lies on the northern horizon. At any other place on Earth you can determine your latitude by measuring the elevation of Polaris in the sky. As I stand on the chalk cliffs at Peacehaven, I am at latitude 50° 47' north of the equator, and about this fact English, French, Germans, Americans, and other peoples of the world had no dispute.

But longitude is an altogether different matter. Where along the equator will we designate the point of zero longitude? Where will we anchor degrees east and west on our maps? The Earth turns under the stars. The stars are no help at all in measuring longitude. One place is as good as any other as a reference point for east and west, and prior to 1884 Britain, France, Germany, and the United States, among other nations, anchored their maps on national observatories. In effect, each nation placed itself at the "center of the world."

The man who forced the issue of standardization was Sandford Fleming, a Scottish immigrant to Canada and a person of irrepressible inventiveness and energy. Before he became an international campaigner for standardized maps and clocks, Fleming had made a name for himself in Canada as a surveyor, mapmaker, and civil engineer. Coming, as he did, from something of a cultural backwater (Canada was then a rather sleepy part of the British Empire), he might seem to have been an unlikely advocate of international standards of longitude and time, but that may be precisely what enabled him to escape the prejudices of local pride. Canada could make little claim for itself as an arbiter of international standards, whereas British and French national pride, respectively, made it almost impossible for one nation to concede to the other primacy of place or time when it came to maps or clocks.

In 1884, at the invitation of U.S. president Chester A. Arthur, and largely as a result of Fleming's ceaseless lobbying, forty-one delegates from twenty-five nations met in Washington, D.C., to decide upon a "prime" meridian, a line of zero longitude that would unify the world's maps. The globe could then be divided into twenty-four time zones, each fifteen degrees of longitude wide, anchored to the prime meridian, and all clocks could keep some hourly multiple of the local time of whatever place was chosen to mark the prime.

To avoid conflicts of national interest, some people advocated a "neutral" prime, anchored on the Great Pyramid at Giza in Egypt, for example, or the temple at Jerusalem, or perhaps the Tower of Pisa in Italy to honor Galileo. These fanciful proposals got nowhere. For one thing, it was important that the prime meridian pass through a first-rate astronomical observatory, as was presently the case for the British and French meridians (the prime prime contenders, so to speak), so that clocks could be kept in sync with the Sun. The Royal Observatory at Greenwich, England, was in many respects the logical choice. Britain had the world's most far-flung empire. Seventy-two percent of international shipping already used maps and clocks based on Greenwich, and railroads in the United States had recently adopted the Greenwich meridian as their basis for standard time. But there was one sticking point: French reluctance to award the prize to their longtime rivals, the British. A French representative to the Washington conference swore that "France will never agree to emblazon on her charts 'degrees west and east of Greenwich'!"

Fleming hoped to allay French pique by proposing an *anti-Greenwich* meridian, a line of zero longitude located exactly halfway around the world from Greenwich, which fell conveniently almost entirely in the watery realm of the Pacific Ocean. This would have the advantage of preserving the integrity of Greenwich-based maps and time zones, without invoking the dreaded G-word on French maps or giving any national capital primacy of place. And the Greenwich observatory could still be used to synchronize time with the Sun. France proposed a compromise: If the English-speaking world would adopt the French meter as the standard measure of length, France would go along with a Greenwich prime meridian. No chance, however, that the British would surrender their beloved traditional standards of measure: inches, feet, yards, miles.

In the end, twenty-two of the twenty-five conference participants agreed to a Greenwich prime meridian, with France and Brazil abstaining and only the tiny Caribbean nation of San Domingo voting against.

And so the peoples of planet Earth took their first partial step toward a concept of space and time that let no person, tribe, or nation claim special privilege. Fleming had in mind something resembling Isaac Newton's universal, absolute space and time, as articulated in the great physicist's 1687 *Principles of Natural Philosophy,* a space and time that made no reference to London, Paris, or Washington, not even to the Earth itself.

Today, with the Internet, geosynchronous satellites, and high-speed air travel, the motions of the Earth around and under the Sun have become increasingly irrelevant. Day and night, summer and winter, are equally suitable for international commerce. The bits and bytes of data that fly through cyberspace at the speed of light take no notice of time zones. In the fraction of a second required for a packet of data to zip from London to Tokyo, the Sun's position in the sky shows no perceptible change. There is no fundamental reason why globally synchronized clocks might not keep a kind of time that makes *no* reference to the Earth's rotation, all terrestrial clocks reading the same hour regardless of the position of the Sun. Hours, days, weeks, months, even years are artifacts of a pretechnological civilization, of increasing irrelevance to a world that never sleeps.

But in 1884 the nations of the world had not yet surrendered their attachment to locality, and even today we cannot claim to be fully citizens of the cosmos. The people of Peacehaven—who in 1984 attached a plaque to their prime meridian monument in celebration of the centenary of the Prime Meridian Conference—clearly take pride in their own perceived place at the center of the world; the Meridian Shopping and Community Centre is the focus of their tidy town—as I discover when I put my back to the sea and begin walking northward along the line of zero longitude.

The walk I intend to make across southeastern England has a purpose; I want to trace humankind's march away from perceived centrality. Singly or collectively, our journey into cosmic space and time begins with a deliberate step, but as I take *my* first step along the meridian I know I have a long way to go before I sever once and for all the psychological umbilical cord that attaches me to the world center of my birth.

For my walk I have equipped myself with maps of the British Ordnance Survey, the national mapping agency, folded sheets from the Explorer series designed for walkers, equestrians, and cyclists. Each sheet covers an area about ten by twenty miles. I have acquired a dozen maps to guide me across southeastern England; they cover the countryside on both sides of the meridian. To celebrate the millennium in the year 2000, the Ordnance Survey issued maps that show the prime meridian as a thick green line, and this will be my approximate route. Of course, it is impossible to walk *exactly* along the meridian—there is no path or right-of-way—but England has an astonishingly (to an American) dense network of public footpaths, all so designated on the maps. It will be possible to make my way from south to north, from the English Channel to the North Sea, without straying more than a few miles from the thick green line. I will walk a good part of that distance during the course of a six-week stay in England.

At Peacehaven, where I begin my trek, my map shows the green line extending out into the sea toward France. The prime meridian is anchored at the Royal Observatory at Greenwich, near central London, but it extends from pole to pole. One hundred miles across the English Channel the meridian comes ashore near Le Havre, France, then passes a bit more than one hundred miles to the west of Paris, where until 1884 the French had anchored their own line of zero longitude. The meridian crosses the Pyrenees into Spain and leaves the Spanish coast at the Gulf of Valencia. Thence down across western Africa: Algeria, Mali, Burkina Faso, Togo, and Ghana. It crosses the equator in the Bight of Guinea and has a watery course across the South Atlantic Ocean until it reaches Antarctica. Going north from Peacehaven, the meridian passes through the Royal Observatory at Greenwich, then follows the valley of the Lee River halfway to Cambridge. It leaves the British coast at Tunstall in Yorkshire and has an unobstructed passage across water and ice to the north pole. The meridian touches nine countries on three continents (if one can call Antarctica a country as well as a continent). Nearly two thirds of the line lies over water.

It is easy to understand why the French, in particular, resisted the adoption of the Greenwich prime. Not only was it *not* the Paris prime; it rather assertively sliced across France, effecting a successful cartographical invasion where centuries of British military interventions had failed. But more to the point, in the 1790s the French had invested a heroic effort and considerable national pride in surveying the Paris meridian from Dunkirk in the north to Barcelona in the south for the purpose of adopting a new national standard of length to be called the meter. The meter was defined by the French National Assembly in 1791 as one ten-millionth of the distance from the Earth's pole to the equator. The idea was to replace an arbitrary profusion of local measures with a national—and hopefully international—standard of length that could be determined precisely by survey and which emphasized the universal brotherhood of man. After all, the one thing shared by all citizens of Earth is the almost perfectly spherical planet they live on. The French quest for universal standards of space and time was a scientific corollary to the Enlightenment idea of universal human rights. It is not coincidental that Thomas Jefferson, who wrote, "We hold these truths to be self-evident, that all men are created equal . . . ," was an early champion of the metric system.

The fraction one ten-millionth of a quarter circumference of the planet was not arbitrary. The size of the Earth was known with reasonable accuracy in 1791, and the French Assembly recognized that one ten-millionth of a quarter circumference would yield a convenient measurement standard, not much different from the English yard. What was now required was to measure the circumference of the Earth with unprecedented precision, incorporating all of the improvements of calculation and instrumentation that accompanied the flowering of Enlightenment science. To this end two prominent astronomers were authorized by the French government to survey a sizable arc of the Earth's circumference. Jean-Baptiste-Joseph Delambre was to travel north to Dunkirk, on the Channel coast, and begin working southward. Pierre-François-André Mechain would journey to Barcelona in Spain and work northward.

The two surveyor-astronomers shared similar backgrounds.

Delambre (1749–1822) was the son of a draper. In the second year of his life he almost lost his eyesight to smallpox. At the age of twenty his vision was so bad that he could hardly read his own handwriting. Nevertheless, he acquired a broad classical education and essentially taught himself mathematics and astronomy. Recognized for his talent, he was taken under the wing of the eminent astronomer Joseph-Jérôme Lefrançais de Lalande, and by 1789 Delambre had established himself as one of the nation's most accomplished theoretical and observational astronomers, in spite of his poor eyesight. Mechain (1744–1804) came from an equally humble background; his father was a plasterer. As a boy he took up astronomy as a hobby, and this led in due time to a distinguished career in science. He too was recognized for his talent by Lalande, who helped Mechain find professional employment as a cartographer. Both men were natural choices for the mission to measure the Earth.

Here is how the two astronomers measured the meridian arc. First, it was required that the latitudes of the end stations in Dunkirk and Barcelona be precisely known, so that one would know exactly what fraction of the Earth's circumference was represented by the measured arc. From hundreds of painstaking observations of Polaris and other stars in the northern sky, Delambre determined the latitude of the Dunkirk starting point to be 51° 2' 6.66". Hundreds of miles to the south, Mechain found the latitude of his Barcelona station to be 41° 21' 45.10". The difference, 9° 40' 21.56", meant that the surveyed distance between Dunkirk and Barcelona was about one thirty-sixth of the circumference of the Earth. Of course, the actual fraction of the circumference was calculated with great exactitude.

Next, the surveyors laid out a long straight line on the ground and measured its length with the most exact "yardstick" available.

From the ends of this baseline they measured with a specially designed instrument the angles formed by the baseline and lines of sight to a distant point observed through a telescope: a mountaintop, tower, or steeple. If one knows the length of one side of a triangle—the baseline—and the adjacent angles, the lengths of the other two sides can be computed, as every high school trigonometry student knows. Any leg of this primary triangle can now stand as the base for another triangle, and another, and another, and so on. The surveyor need only move his instrument from vertex to vertex, measuring angles, to extend a web of triangles across the countryside, and thereby ultimately determine the distance of any vertex of the web from any other. This Delambre and Mechain did, along an approximately ten-degree arc of a north-south meridian, or one thirty-sixth of the circumference of the Earth. Heavy instruments had to be carried to each vertex—hilltops, mountaintops, or towers—leveled, made steady, angles measured and measured again, data recorded, checked, and rechecked. Like navigators at sea, Delambre and Mechain contended with cloudy skies. They sometimes waited for days on windy, cloud-socked mountains before they could spy a distant beacon through the telescopes of their angle-measuring instruments. The web they painstakingly stretched across the length of France and part of Spain required hundreds of interlinked triangles, surveyed from north to south and south to north, with vertices in high places that afforded the widest possible view of the countryside, until at last the two surveyors, with their crews, met somewhere south of Paris and married their nets of triangles.

The surveyors' goal, remember, was not simply to know the size of the Earth; they wished to know the size of the Earth *as exactly as possible* so that the world might be provided with a new standard measure of length, the meter. Meanwhile, France was in turmoil. That the two men managed to complete their survey in the midst of a popular revolution and war with Spain, without losing their heads to the guillotine or their instruments to brigands, is little short of astonishing. Mechain especially suffered trials and tribulations that would have broken a lesser man, not least inconsistencies in his data for the latitude of Barcelona that troubled him grievously until the end of his life. Both men were exceptionally devoted to the ideals of science. When Delambre presented the final report on the meridian survey to Napoleon in 1806, the emperor said: "Conquests will come and go but this work will live forever." Mission completed, a platinum bar was contrived to the requisite dimension—the first "meter stick" against which all other length-measuring instruments would henceforward be calibrated. The Earth's size had been determined more accurately than ever before, and the peoples of the Earth (or at least of France) had a new system of measurement.

Thomas Jefferson was solicitous that the new nation of the United States of America embrace the French metric system, but he was not so optimistic as to imagine that it might actually happen. To this day, the United States is almost the last holdout in its official adherence to feet, miles, gallons, and pounds. Americans are now in the same position with regard to the metric system as were the French with regard to the Greenwich meridian; they jingoistically thumb their noses at the rest of the world, even at the expense of their own best economic interest.

When I was a child I learned to measure with foot-long rulers and yardsticks, the latter usually obtained free at the local hardware store and printed with advertising. I came across my first meter stick in high school physics lab. It was only when I went off to study science at university that I became aware that a platinum bar in Paris ruled the world of scientific measurement. Since then I have done all my quantitative thinking about the world in metric, but whenever I write for a popular American audience my editors insist that I express all dimensions in feet and miles, as I have done in this book. Americans, it seems, are loath to surrender their self-proclaimed role as arbiter of the world's standards.

For a brief period in the 1970s the United States *almost* adopted the metric system, but popular resistance was fierce and the proposed realignment of measures was put aside by President Ronald Reagan when he came to office. Meanwhile, global trade has exerted an inexorable pressure for change, and American industry—the automotive industry in particular—has transformed itself to metric even in the face of official intransigence. Eventually, of course, policy will change to reflect practical reality. The thrust of history is always away from local assertions of centrality.

In fact, even the platinum bar in Paris has lost its prominence. Since 1983, the meter has been defined as the distance traveled by light in a vacuum during a time interval of precisely 1/299,792,458 second. Likewise, the second is no longer defined as a fraction of the rotational period of the Earth (which is slowing down), but as the duration of 9,192,631,770 vibrations of the radiation corresponding to the transition between the two hyperfine levels of the ground state of the cesium 133 atom at rest at zero degrees absolute temperature. (Don't worry about what that last sentence means; suffice it to say that atomic vibrations can be *very* precisely measured by physicists.) For the time being at least, physicists consider the speed of light and the oscillation frequencies of atoms to be constant throughout the universe, and therefore the revised metric standard is presumably independent of *any* nation's claim of precedence. The new definitions make no reference to the Earth at all.

Delambre and Mechain were not the first to measure the Earth. Greek natural philosophers of classical times understood that the Earth was a sphere and computed its size with astronomical observations. The person credited with the first scientific determination of the Earth's circumference is Eratosthenes (c. 276–196 B.C.), librarian of a great repository of books at the city of Alexandria at the mouth of the Nile River in Egypt. We know almost nothing about Eratosthenes, other than that he hailed from Cyrene, a city to the west of Alexandria on the north coast of Africa, and lived in Alexandria after the death of Alexander the Great and before the rise of Rome as a great power. We know nothing of what he looked like, his ancestry, his progeny, his virtues, or his vices. We do know that he determined the size of the Earth and did so with surprising accuracy.

When Alexander's armies conquered Egypt in the fourth century B.C., he established there a city of his own name that soon became the philosophical capital of the Mediterranean world, an intellectual mecca that drew to its shining white streets philosophical and mathematical luminaries from throughout the Mediterranean world. The city possessed the finest library in the world, with tens of thousands of scrolls, over which Eratosthenes presided. According to some sources, Eratosthenes was scorned by a number of his contemporaries as a jack-of-all-trades and master of none, but this may be an ideal qualification for a librarian.

Exactly how Eratosthenes made the epic discovery that ensured his fame we will never know; we have only the sketchiest clues from historical sources. I like to imagine it happened something like this: A man of unquenchable curiosity, Eratosthenes haunted the marketplaces and docks of Alexandria, quizzing caravanners and sailors about the geographies and cultures of the places they had visited. One day, he heard from a caravanner of a deep well in the town of Syene, some distance down the Nile (near where the Aswan Dam stands today), where on a midsummer day the Sun can be seen reflected in the water at the bottom of the well. This implied that on that day the Sun was directly overhead at Syene. At first, Eratosthenes tucked this trivial fact away in his capacious memory, but later that evening (I continue my fantasy), after a few drafts of wine perhaps, he leaped to his feet and exclaimed to his friends, "I know the size of the Earth!" To which announcement his friends no doubt exchanged winks and nudges and ordered another round of drinks. But Eratosthenes knew (because he was, after all, a jack-of-all-trades) that the Sun was *not* overhead on a midsummer day in Alexandria. And he knew why: The Earth is a sphere. If the Sun is far away and its rays are essentially parallel, sunlight will fall at different angles on different parts of the Earth's surface. (See figure 1-1.) If the Sun is directly overhead at Syene, it cannot be directly overhead at Alexandria. This was the essential clue that gave the game away.

On the next midsummer day, Eratosthenes made a simple quantitative observation. He measured the length of the shadow of a vertical column at Alexandria, and therefore the angle of the Sun's rays with the column. This, with the knowledge of the well in Syene, was the equivalent of Delambre and Mechain measuring the difference in latitude at each end of their surveyed line between Dunkirk and Barcelona. The angle Eratosthenes measured at Alexandria when the Sun was at Syene's zenith was a bit more than seven degrees, or about one fiftieth of a circle, and this he

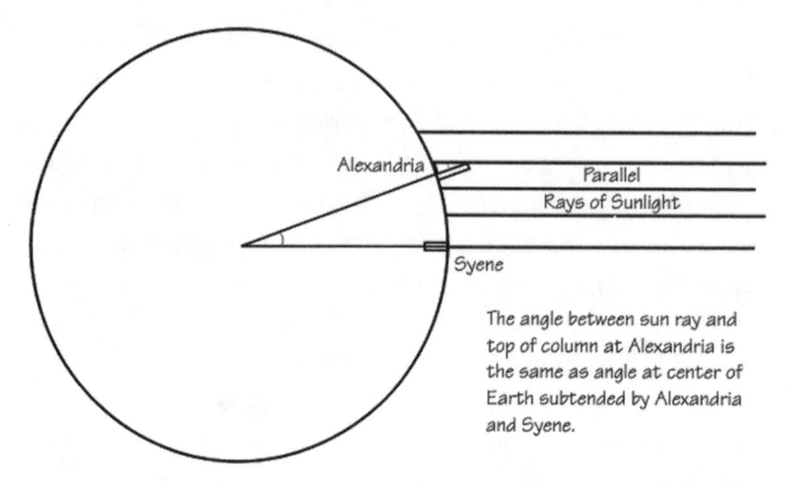

FIGURE 1-1. Eratosthenes' method for measuring the circumference of the Earth.

knew must be the difference in latitude between Alexandria and Syene (although, of course, no well-defined notion of latitude yet existed). In other words, the distance from Alexandria to Syene was about one fiftieth of the circumference of the Earth. Eratosthenes was in no position to carry out a survey of the distance from Alexandria to Syene, which was hundreds of miles away down the valley of the Nile, in the fashion of Delambre and Mechain. But he did know that this was a journey of about fifty days by caravan, and that caravans traveled about 100 stadia a day (a stadium, the Greek measure of distance, was apparently about one tenth of our mile). The distance between the two places is therefore approximately 5,000 stadia, and this corresponds to one fiftieth of Earth's circumference. The Earth's circumference is therefore 250,000 stadia, Eratosthenes determined—about 25,000 of our miles—a measure that is remarkably accurate given the roughness of his data.

This story is often recounted in books on the history of

astronomy or cartography, as well it should be, but it seems to me that the two most striking aspects of the story are seldom mentioned.

First, look again at the diagram (figure 1-1) so often produced to illustrate Eratosthenes' method. The Earth is represented by a circle. This seems completely natural to us; after all, we have seen photographs of the Earth from space, as round and smooth as a billiard ball. But up close, here on the surface, the Earth looks like anything *but* a perfect geometric sphere. Plants, animals, hills, valleys, rivers, seas, cities, temples, waves, clouds: All of this—all of the intricate diversity of the Earth, everything that is of interest and important to the lives of men and women—Eratosthenes dismisses in his diagram. He draws a circle with a compass and says, "This is the Earth." Then, having reduced our multivarious globe to a pure geometric figure, he calculates its size. This, I maintain, was a pivotal moment in human history, an act of stupendous intellectual abstraction that stands as the beginning of mathematical science.

Here for the first time we see the three pillars of scientific method working together: (1) an idealized conceptual model of the world (the Earth as a perfect sphere); (2) quantitative observation (measuring the angle of the shadow and the distance from Alexandria to Syene); and (3) mathematical computation (in this case, the rules of Euclidian geometry). Greek civilization gave us many wonderful things; try to imagine Western civilization without the underpinnings of Greek politics, art, architecture, drama, history. But Eratosthenes' drawing of the Earth as a geometric circle represents something as formidable as the plays of Sophocles, the history of Herodotus, or Athenian democracy: a way of abstract thinking that would eventually carry human imagination to the far-off galaxies.

Second, something more subtle, almost never commented upon: I checked half a dozen astronomy books; they all have a version of figure 1-1, with the Sun's rays falling parallel onto the Earth. But this can only be so if the Sun is *very far away* compared to the size of the Earth. If not, the rays will not be parallel and Eratosthenes' demonstration falls apart. Indeed, Eratosthenes might equally well have assumed that the Earth is flat and used his observations of shadows and distances to calculate the distance to the Sun. (See figure 1-2.) Implicit in the innocent-looking diagram of the astronomy books is an intuition on the part of Eratosthenes that is audacious in its daring: *The Earth is a tiny sphere compared to the distance to the Sun. And if the Sun is far away, it must be very large, perhaps even larger than the Earth.* Here is a guess by Eratosthenes upon which all depends—the smallness of the Earth compared to the distance to the Sun—something we know with certainty today but which in Eratosthenes' time could not be proved and indeed violated common sense. Why would Eratosthenes assume such a thing? An inspired guess? Perhaps. But as we shall see in the next chapter, Alexandrian scientists (the first

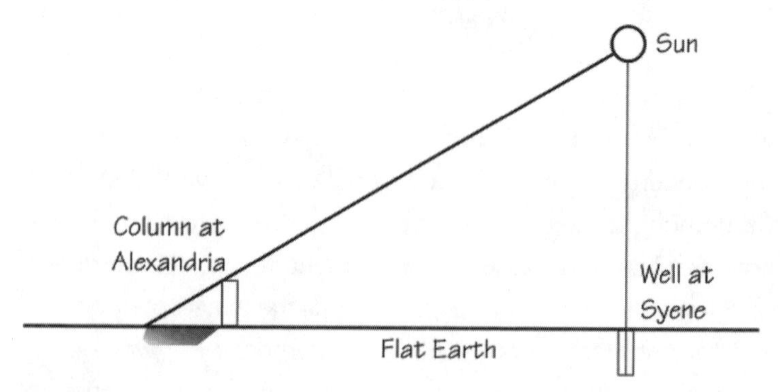

FIGURE 1-2. An alternate interpretation of Eratosthenes' observations, assuming a flat Earth.

real scientists) were thinking not just about the size of the Earth but also about the size and distances of the Sun, Moon, and stars, about eclipses and the phases of the Moon, about the way the positions of celestial bodies change as one moves north and south on the surface of the Earth. Eratosthenes' measure of the Earth was part of an evolving network of new astronomical ideas. At work is what we might today call *economy of explanation*: We hold to be true those hypotheses that explain the most in terms of the least. No divine revelation confirmed for Eratosthenes the truth of his core beliefs. All science rests on a foundation of inspired guesses. What matters is not that we can prove the foundational assumptions of scientific explanation, but *consistency and economy* in an overall scheme of explanation.

From the prospect of the chalky highlands above Peacehaven the sea recedes to the far horizon. At the line where gray water meets blue sky I can just make out the black bulks of freighters and tankers making their way up-channel. Beyond the horizon lies the France of Delambre and Mechain, and even farther afield the Mediterranean world of Eratosthenes. I try to imagine standing in this same place in Eratosthenes' time, when Rome, Athens, and Alexandria were the glittering centers of Western civilization and England lay on the barbaric and crudely sketched border of their maps. But Peacehaven and Alexandria have one thing in common: Across the centuries they have each laid claim to place on a prime meridian. Peacehaven has had that privilege since 1884; Alexandria had its day in the cartographic sun during the several centuries before the Christian era.

Alexander founded Alexandria after a dream in which a

venerable prophet indicated to him the site of the city. It is said that Alexander is buried in Alexandria, but no trace of his tomb remains. The city was built on a ribbon of land between Lake Mareotis and the Mediterranean Sea, with forums, temples, marketplaces, palaces, a double harbor with a famous lighthouse, quays, warehouses, and, prominently, a museum (place of the muses) and the famous library over which Eratosthenes presided. The museum and library together were the equivalent of a great modern university. It was the dream of the first rulers of Alexandria (the Ptolemys) that the library would possess a copy of every book in the known world, and within a century hundreds of thousands of scrolls were collected within its walls. By the middle of the first century B.C., Diodorus of Sicily could say that Alexandria was "the first city of the civilized world, certainly far ahead of all the rest in elegance and extent and riches and luxury."

Alexandria was culturally a Greek city, an anomalous Hellenic appendage to Egypt. Native Egyptians and Egyptian culture never achieved more than second-class status among the city's shining thoroughfares. Intellectual traditions that had their genesis in fifth- and fourth-century Athens with Plato and Aristotle (Alexander's teacher) moved across the sea to the north shore of Africa and took root in the spanking-new metropolis. Here was an atmosphere that welcomed all comers—Eratosthenes from Cyrene, Aristarchus from Samos, Archimedes from Sicily, Apollonius from Rhodes, Hipparchus from Nicaea, Galen from Pergamon, and so on—the only requirement being, apparently, an inquisitive mind and a bent for explaining the world in terms that made no reference to the gods. Mathematics, astronomy, geography, mechanics, and medicine reached levels of development that would not be surpassed for nearly two thousand years.

Two questions come to mind: What sparked such a fruitful

commotion of ideas? And why, after a few centuries, did the bright flame of Alexandrian science fizzle out?

The city's location partly answers the first question. Alexandria could hardly have been more advantageously placed to become a commercial and economic powerhouse, situated as it was at the nexus of three continents and connected by canal to the greatest river of the known world, the Nile. With trade of material goods goes a corresponding flow of ideas. Any book that entered Alexandria's harbor aboard a ship was required by the local authorities to be copied; the original went to the library and the copy was returned to the original owner. Moreover, a culture built on commerce is fertile ground for science; there is a mutually advantageous correspondence between the entrepreneurial spirit and science's questing openness to new ideas and fondness for innovation. A polycultural mix of ideas came felicitously together at Alexandria. From Greece came the gift of philosophical abstraction. From seafaring traders and from Egypt itself came a firm interest in the material *stuff* of the world: earth, water, air, fire. The abstract and the practical. Mathematics and technology. These were the metal and the flint that ignited the spark of Alexandrian genius, a combination of influences that would not be repeated until the time of the European Renaissance.

These two pillars of Alexandrian science—abstraction and practicality—were soon enough wrenched apart. First came Rome, with its hard-nosed focus on technology: roads, aqueducts, sanitation, military power, and other accoutrements of empire. The Romans had no gift for abstraction. Their genius was pragmatic, matter-of-fact. Then came Christianity, with its born-again attachment to a spiritual otherworld. These diametrically opposed cultural forces swept through Alexandria in recurring waves, often with violent and destructive consequences. The happy marriage of abstraction and empiricism that animated Alexandrian science fell upon troubled times.

But for a few centuries in this exceptional place human imagination soared beyond the local horizon into apparently limitless cosmic spaces and unseen parts of the terrestrial globe. Not only did Eratosthenes measure the globe; he mapped it. (See figure 1-3.) Alexandria was an ideal place to gather information from travelers to all parts of the known world, and Eratosthenes was apparently voracious in his interrogations. As far as we know, he was the first to contrive a map of the known world on a north-south, east-west grid, like our present system of meridians and parallels.

As we have seen, it is easy to locate a place's north-south position on a map by noting the angular elevation of celestial bodies—such as the Sun or Polaris—above the horizon. But the sky provides no clue as to one's east-west position on the globe. The best that Eratosthenes could do was ask mariners and caravanners to estimate the distances to places they had traveled. His map reached from the Iberian Peninsula in the west to India in the east, with meridians drawn at prominent geographic features: the Straits of Gibraltar, Carthage, the mouths of the Red Sea and Persian Gulf, the Indus River. The map was most accurate in the regions near Alexandria; such faraway places as Britain and India were known only sketchily. If there was something that could be called a prime meridian on Eratosthenes' map, it was the vertical line near the center of his map that went through Alexandria and down along the Nile River to Syene. The Alexandrian meridian was to Eratosthenes what the Paris meridian was to Delambre and Mechain.

The Alexandrian astronomer Hipparchus, who was born just about the time of Eratosthenes' death, professed himself deeply

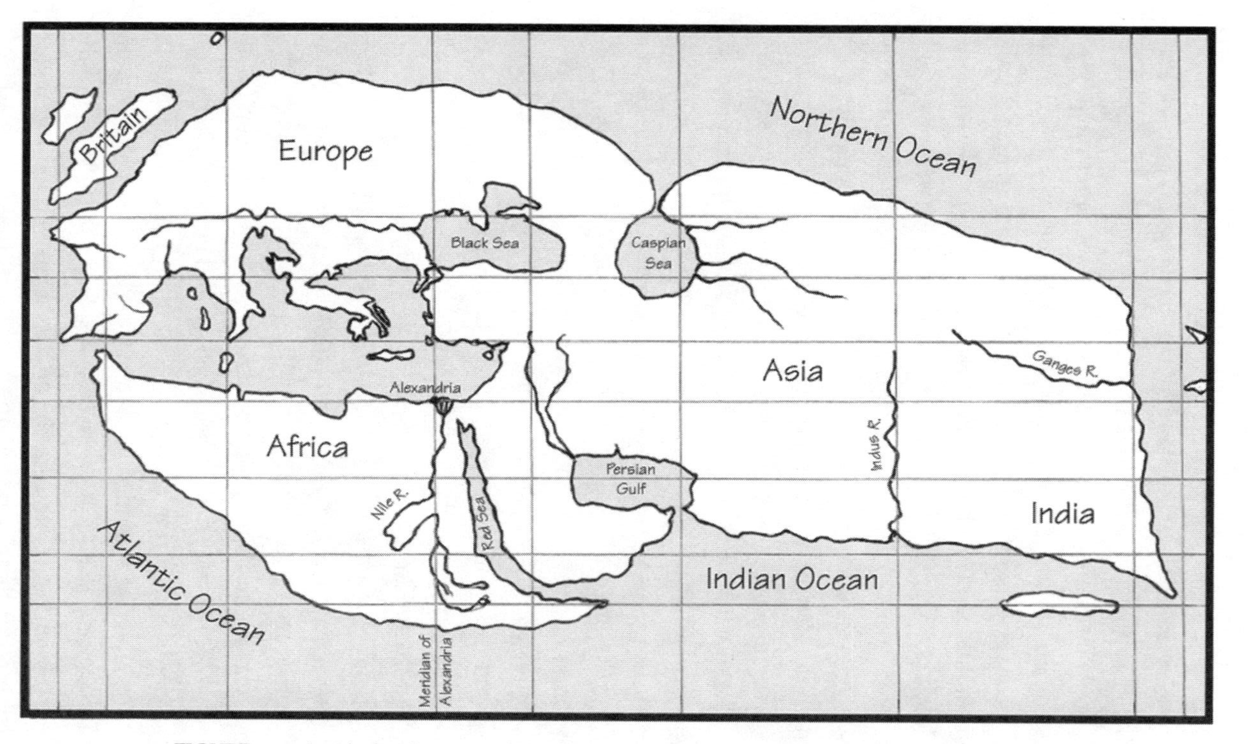

FIGURE 1-3. An idealized version of Eratosthenes' map of the world, with modern nomenclature.

disappointed in his predecessor's map. He wrote an angry tract, *Against Eratosthenes,* taking the earlier geographer to task for the slapdash appearance of his geographic grid. Meridians and parallels should be established at equal intervals on the terrestrial globe, insisted Hipparchus, just as astronomers do in the sky. Hipparchus was a consummate theoretician, with a passion for mathematical representations of reality. For example, he was the first, as far as we know, to quantify the brightnesses of stars; when today's astronomers speak of the "magnitudes" of stars—first, second, third, and so on—they are using Hipparchus's scale. Of course, Eratosthenes was no mathematical slouch himself, and he had undoubtedly recognized the value of a uniformly spaced grid of meridians and parallels, but he also knew that it didn't make much sense to have a mathematically precise system of longitude when the actual locations of places were so imprecisely known.

Three hundred years after Eratosthenes and Hipparchus, the Alexandrian geographer and astronomer Claudius Ptolemy (second century A.D.) compiled an atlas of the world that stood unsurpassed until the European Age of Exploration began in the sixteenth century A.D. Ptolemy's map of the world (figure 1-4) adopted the uniform grid proposed by Hipparchus, and Ptolemy was the first to make a stab at representing the sphericity of the Earth on a flat piece of paper. His map was impressively accurate with respect to latitude, but no reliable method of obtaining longitude was yet available. Ptolemy placed his prime meridian (zero degrees of longitude) at the westernmost land he knew about, the Fortunate Islands in the Atlantic Ocean (the present-day Canary Islands). His map extended eastward to what we would now call Southeast Asia, and Ptolemy believed this represented about half the circumference of the Earth, or 180 degrees. Unfortunately, Ptolemy seriously underestimated the size of the Earth. He took

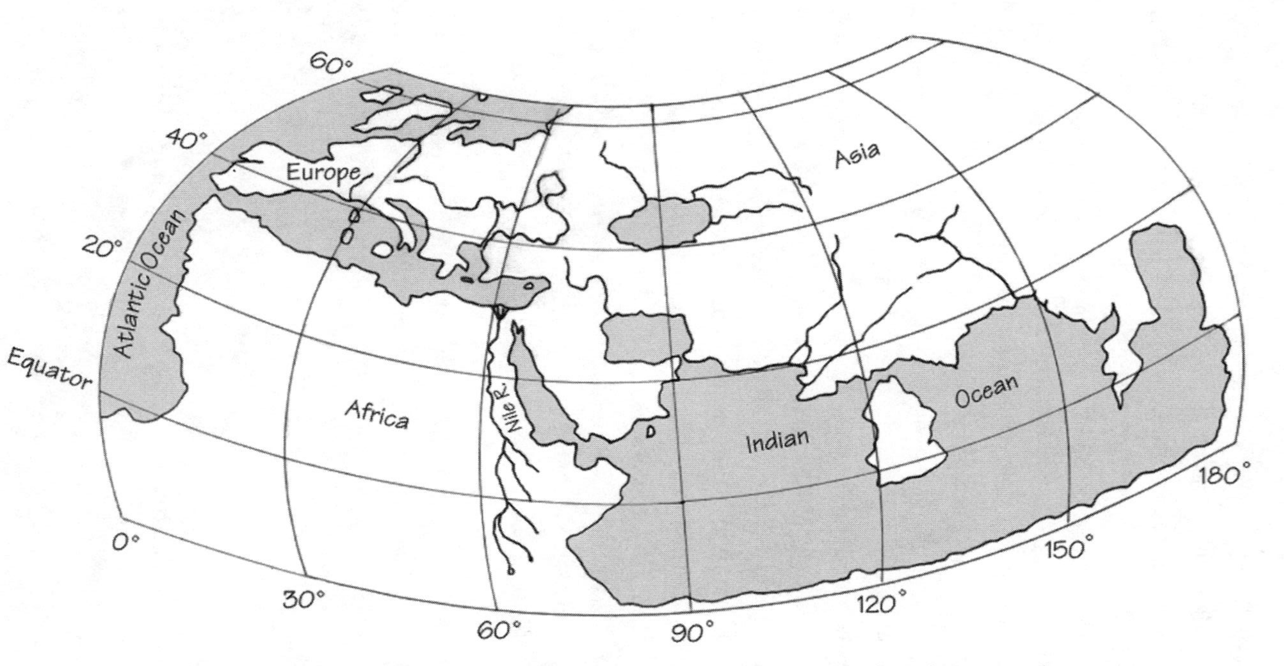

FIGURE 1-4. An idealized version of Claudius Ptolemy's map of the world, with modern nomenclature.

the diameter to be 18,000 miles, rather than the more accurate 25,000 miles of Eratosthenes. Or perhaps I should say "fortunately," since it was the influence of Ptolemy's atlas that later inspired Columbus to try to reach the Indies by sailing westward. Following Ptolemy, he had every reason to believe that Asia was no farther away to the west than to the east. When he bumped up against the Americas, he believed he had reached his goal. Had the Genoese navigator comprehended the actual size of the Earth, he might have been less willing to launch himself into the wild, blue deep.

Claudius Ptolemy was the last of the influential geographers and astronomers of Alexandria. Disruptive forces were afoot in the Mediterranean world. Ethnic and religious strife abounded. Dogma triumphed over tolerance. Theological speculation overwhelmed scientific inquiry. Presumed miracles now inspired more interest than the ordered celestial events that had been the focus of Alexandrian science. Sometime between the Roman conquest of Egypt in the first century B.C. and the Muslim conquests of the seventh century A.D., perhaps in several episodes of vandalism, the great library went up in smoke—and the world's greatest repository of knowledge was sacrificed to the frenzied passions of true belief. When Europe rediscovered Ptolemy's maps at the time of the Renaissance, the prime meridian was still attached to the Fortunate Islands.

Ptolemy's map of the world depicted about a third of what he presumed to be the Earth's surface, from longitude zero at the Fortunate Islands to longitude 180 degrees somewhere in Southeast Asia, and from 15 degrees south of the equator to 65 degrees north of the equator. He knew therefore that the greater part of the world was beyond his ken, and this knowledge, restored at the time of Europe's Renaissance, inspired sailors to embark upon journeys of discovery into the blank spaces of the map. This is the power of the scientific method invented by the Alexandrians: It forces us to look beyond the familiar.

Each of our lives is to a greater or lesser extent a journey into the "blank spaces of the map." Inevitably, unconsciously, we begin that journey during the first years of our lives, from an egocentric world of direct sense impressions into a space that extends beyond direct perception, into a past we never experienced, and into a future of unrealized possibilities. We have no memory of this early stage of our intellectual development, and so I cannot tell you how it was I first came to know that the world accommodated more than *this* place and *this* moment. How much of that early intellectual development is universal—that is, how much is nature rather than nurture—is hotly debated by child psychologists, anthropologists, and linguists. My own intellectual view of the matter was largely shaped by the empirical researches of the Swiss child psychologist Jean Piaget, who during the mid-twentieth century published a number of influential books on the inner lives of children.

I came across these books as a young college professor keenly interested in the history of science. I was struck by how faithfully the intellectual development of very young children, as studied by Piaget, paralleled the early history of science. In both cases, the movement is from egocentrism to relatively uncentered space and time. In both cases, the earliest understanding of the world is animistic (everything is alive) and artificialist (everything is the result of conscious agency). The child who puts a happy face on a crayon drawing of the Sun is not being merely metaphorical, at least not initially; very young children truly imagine that the Sun is in some sense alive. ("The Sun follows me when I walk.") Likewise, ask a child *why* the Sun is in the sky, and the answer will be some version of "It was put there for me." Anthropologists have shown us that human thinking is initially animistic and artificialist among all peoples of the Earth, and huge residues of animistic and artificialist thinking remain with us even in an age of science. The parallel between the intellectual development of children and of the history of science was not lost on Piaget, who had begun his career as a biologist.

According to Piaget, the young child does not easily separate the thought of a thing from the thing itself. There is for the child an instantaneous and spontaneous tendency to confuse internal and external worlds, the psychical and the physical. Only with time does the child come to recognize the *external* and *independent* reality of Sun, Moon, wind, clouds, trees, stars. With maturation, the child recognizes a space and time that is independent of her own perceptual universe. Piaget believed these stages of development are universal and innate, and that a child must be allowed to progress through a construction of reality at her own pace. The progress of science away from animism and artificialism is a logical extension of the child's intellectual development, thought Piaget. Animism and artificialism are projections of self onto a world that exists independently of ourselves. External reality, however imperfectly known, is the touchstone by which we can measure individual and cultural maturity.

It has become fashionable of late among certain academics to suggest that the child's earliest understanding of the world—as animistic and artificialist as it might be—is no less legitimate than the world described by science, and that the "development" Piaget speaks of is actually a kind of brainwashing of children into the worldview of adults. These academics are equally likely to stress the cognitive equivalence of all cultural forms of adult knowledge; the cosmology of a Stone Age people of New Guinea, say, is a no less valid representation of reality than modern science. There is no knowledge of the world that is independent of the knower, insist these modern relativists, and the knowledge enshrined in so-called Western science has no more claim to objectivity than any other kind of knowledge.

This sort of relativism does not measure up against the story I have told of Alexandrian science. No one questions that what happened at Alexandria during the last centuries of the pre-Christian era was culturally conditioned. We have seen how a unique combination of theoretical abstraction and practical empiricism came fortuitously together to create a new way of knowing and new knowledge of the world. However, it requires a perverse attachment to relativism to deny an objective reality to the world constructed by Eratosthenes, Ptolemy, and their contemporaries. After all, we have seen and photographed the spherical Earth from space. We have sent spaceships around the Sun and bounced radar off the planets. We have confirmed and refined the cosmological speculations of the Alexandrians. No scientist can *prove* that a physical world exists "out there" that is independent of our knowledge of it, nor can we prove that science can know the world in an ever more objective way. But without the *assumption* of objectivity there is no motive for pursuing science. Without the *assumption* of objectivity the geometric diagram of Eratosthenes, by which he measured the Earth, has no more significance than the smiley face on a child's drawing of the Sun. Look again at figure 1-1. Clearly there is a distinction between the sign (the circle) and the thing signified (the Earth), and Eratosthenes was clearly aware of the distinction. But he was surely confident that his diagram had objective correlates in reality, and that mathematical deductions based on the diagram, like his maps of the known world, added to humankind's store of reliable knowledge.

Jean Piaget was a child prodigy who became interested in science at an early age. He wrote and published his first scientific paper at the age of ten, a short note on his sighting of an albino sparrow. He honed his observational skills with studies of the mollusks of Lake Geneva and earned a Ph.D. in zoology. Eventually he turned to the study of the inner life of children precisely to gain a deeper understanding of the scientific quest for reliable knowledge of the world. It is undoubtedly comforting for a child to imagine a Sun who smiles and follows; it requires rather more daring to accept a Sun that is vastly larger than the Earth, a sphere of fire, inorganic and indifferent. The Sun we study with satellite telescopes today is a far cry from the humanlike Sun god of the Egyptians or the Greek Helios who each day drives his golden chariot across the sky. Knowledge is *not* all relative. Some knowledge is more reliable than others. To accept this possibility is called growing up.

2

THE EARTH IN SPACE

The path up into the South Downs from Peacehaven provides splendid views back across the English Channel. The day of my walk is clear, not a cloud in the sky. At no point along my path, however, is the coast of France visible. The continent of Europe is hidden out there behind the curve of the Earth. From my lofty perch on the chalky uplands the gray water of the Channel seems to extend forever, and I marvel at the courage of the men and women of the past who dared to set out in small craft to see what lay beyond the apparent infinity of a sea horizon.

On this October day, the Sun is high in the south, and in the west a waning Moon droops low in the sky. I stretch out my arm to the south and then to the west and measure the apparent sizes of the Sun and Moon against my little finger: Both celestial objects appear *the same size* in the sky, about half a finger's width at arm's length. According to modern astronomers, this equality of the *apparent* diameters of Sun and Moon is a coincidence, a cosmic accident. The Sun just happens to be four hundred times farther away than the Moon and four hundred times as wide. The coincidence has consequences, however: If the Moon were farther away from us relative to the Sun, we would not have total solar eclipses, because at the greater distance the Moon's smaller size in the sky would not cover the Sun's face; if the Moon were closer, its apparent size would be greater and solar eclipses would be less rare.

Alexandrian astronomers assumed the Moon was closer than the Sun and lit by the Sun's light, because doing so made possible elegant and economical explanations of solar eclipses, lunar eclipses, and the changing phases of the Moon. The Alexandrian explanations are deemed correct today. But how far away are the Sun and Moon? And what are their true sizes compared to the Earth? These questions too the ever-ingenious Alexandrians answered, particularly Aristarchus (c. 310–230 B.C.).

During the year I spent studying the history of science at London's Imperial College, I came across Sir Thomas Heath's translation of Aristarchus's *Treatise on the Sizes and Distances of the Sun and Moon.* It was an eye-opener, a tour de force, to my mind the most compelling work of science we have from the ancient world. So perfectly does the book illustrate the power of the scientific method invented at Alexandria during the third century B.C. that I subsequently made Aristarchus's little volume on the architecture of the cosmos a centerpiece of my teaching. The book itself is not an easy read—it is written in the language of Greek mathematics and profusely illustrated with complex geometric diagrams—so some simplification is necessary when I guide students through it, but what Aristarchus describes in his book is not beyond the understanding of any present-day high school student.

However, Aristarchus's discoveries were so radical *for his time* that more than seventeen centuries would pass before substantial numbers of people were prepared to accept his view of the universe. It seems a miracle that his book survived at all. Of course, no copy from his own time is extant; the earliest existing manuscript of the text dates from the tenth century. The book has come down to us through a process of copying and recopying, often, presumably, by scribes who did not themselves fully understand what they copied or grasp its significance.

Here, for instance, is Proposition 15 of the book, which Aristarchus proves by invoking his previously asserted axioms and proved theorems: *The diameter of the Sun has to the diameter of the Earth a ratio greater than that which 19 has to 3, but less than that which 43 has to 6.* Or, put more simply, the Sun's diameter is six or seven times greater than the Earth's. I try to imagine what Aristarchus's contemporaries made of this conclusion. I stand on the summit ridge of England's South Downs. To the north the vale of the Weald stretches away to the distant North Downs. To the south the sea seems to touch infinity. I have walked all morning, and I can still see the place where I began, the sprawl of Brighton on the chalky coast. I have flown thousands of miles across an ocean to make this walk, to traverse on foot this tiny part of the Earth's surface. In the sky I see the orb of the Sun. I can cover it with the tip of my little finger stretched at arm's length. Yet this tiny glowing disk, Aristarchus tells us, is six or seven times larger than the Earth! Or to give his conclusion a more concrete representation: If the Earth were represented by a grape, the Sun would be a melon. Put down the book you are reading, step outside and look at the Earth sprawling about you and at the Sun in the sky, and try to imagine that the latter is six or seven times larger than the former. This conclusion is so counterintuitive, so seemingly at odds with common sense, that one can understand why it took so long to be generally accepted.

Aristarchus was born on the island of Samos in the Aegean Sea near the coast of what is now Turkey, sometime around 310 B.C. Like many other curious and well-educated persons of his time, he made his way to Alexandria. We know from other sources that he wrote on vision, light, and color, but those works have not survived. We have his book *Treatise on the Sizes and Distances of the Sun and Moon,* but not much else is known about him, so there is ample room for fantasy.

I like to imagine Aristarchus holding forth in, say, a colonnaded courtyard of Alexandria, regaling his colleagues and students with the results of his astronomical investigations. Between the thumb and forefinger of one hand he holds a grape—"Imagine that this grape is the Earth," he says. In the palm of the other hand he holds a melon: "The Sun." His listeners have previously struggled to comprehend what Eratosthenes has told them about the size of the Earth: that the entire known world, from the Atlantic shores of Europe and North Africa to India, is but a fraction of the Earth's surface. Now Aristarchus is telling them that the entire globe of the Earth is minuscule compared to the Sun.

"Hold this," he says, and he hands the grape to one of his audience. Then he backs away with the melon on his outstretched palm. "Tell me," he says, "when the melon has the same apparent size as the Sun in the sky" (that is, half the size of the little finger of an outstretched arm). He is all the way across the courtyard when the person with the outstretched arm says, "Stop." Aristarchus is silent as he lets the implications of his little demonstration sink in. For his audience, the universe has suddenly become almost unimaginably large.

How did he do it? How did Aristarchus deduce the size of the Sun relative to the Earth? First, he imagined a triangle whose vertices are the Earth, Moon, and Sun *at the moment when the Moon appears exactly half full in the daytime sky.* (See figure 2-1.) The angle at the Moon must be a right angle, Aristarchus reasoned, when exactly half of the Moon's face as seen from Earth is illuminated by the Sun. With some sort of instrument, he then measured the angle subtended at his eye (the Earth) between the Sun and a half-full

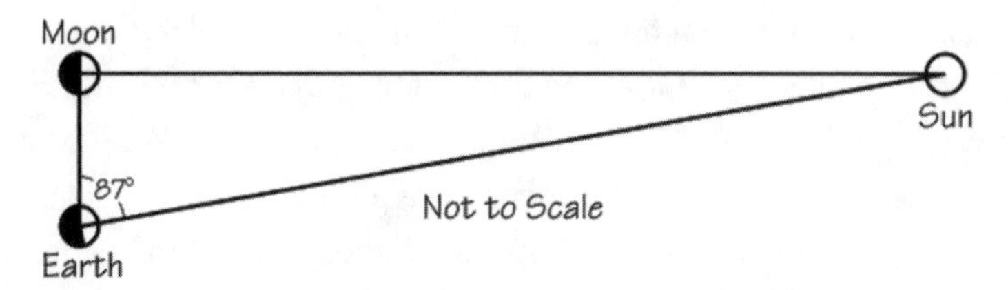

FIGURE 2-1. Aristarchus's understanding of the Earth, Moon, and Sun when the Moon is half full as seen from Earth. Not to scale.

Moon and found it to be about 87 degrees. You can try it yourself; the next time you see a half-full Moon in the daytime sky, stretch out your arms so that one arm points to the Moon and the other to the Sun; you will see that your arms make approximately a right angle to each other. You will not, of course, be able to measure the angle exactly; that would require some sort of angle-measuring instrument such as the one used by Aristarchus, and to be truthful, even Aristarchus had a tough time getting it right. The rest of his demonstration you can also do yourself. Use a protractor to draw a right triangle with an 87-degree angle at one vertex, and label the angles Earth, Moon, and Sun. The other acute angle (the one at the Sun) will of course be 3 degrees. (The sum of the angles of any triangle is 180 degrees.) Your drawing will look like figure 2-1, but the triangle will be much longer and skinnier. Now measure the sides. You will see that side Earth-Sun is nineteen times longer than side Earth-Moon. It doesn't matter how large or small a triangle you draw; all such triangles are similar, and the ratio of their sides will be the same. Of course, the triangle you have drawn on paper is also similar to the triangle of *cosmic dimensions* whose vertices are the *actual* Earth, Moon, and Sun. And therefore the Sun must be nineteen times as far away as the Moon.

And since the Sun and Moon appear the same size in the sky—half the width of your little finger held at arm's length—the Sun must also be nineteen times bigger than the Moon. (You may recall that earlier I said that the Sun is four hundred times farther away than the Moon and four hundred times as wide. We'll come back to this apparent inconsistency in a moment.)

Now you must admit that this is a rather astonishing conclusion from so simple an observation (that 87-degree angle), a theorem of Greek geometry (the law of similar triangles), and a little fiddling around on paper with a protractor and straightedge. You have begun to glimpse the power of the new way of thinking invented by the Alexandrians.

But how far away are the Sun and Moon? And how big are those bodies compared to the Earth, whose staggering dimension Eratosthenes had measured? What comes next is a dazzling piece of creative thinking, and to recreate it is to have a better understanding of the power of scientific methodology and the achievement of the Alexandrians.

First, note with Aristarchus that both Sun and Moon subtend an angle of about one half a degree in the sky (half a finger breadth at arm's length), which means they are about 115 times as far away as they are wide. Try it. Measure half the width of your little finger and the distance of the fingertip from your eye with arm outstretched and take the ratio. You should get something close to 115.

Next, Aristarchus observed a total eclipse of the Moon, a not uncommon celestial event. Like his colleagues, he believed the Moon shines by the Sun's reflected light and that it goes dark during a lunar eclipse because it moves into Earth's shadow. (See figure 2-2.) He timed the eclipse, perhaps with a sand glass, and observed that the Moon remains dark for twice the time it takes to enter into or emerge from the shadow. This means that the Earth's shadow at the distance of the Moon must be twice as wide as the Moon. From his earlier observations and calculations

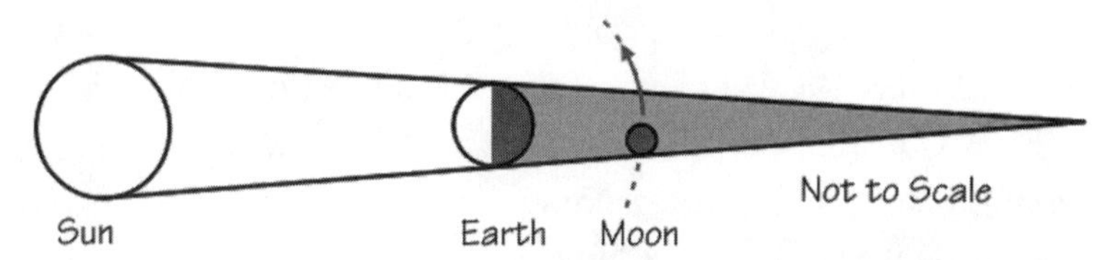

FIGURE 2-2. Aristarchus's understanding of Earth, Moon, and Sun during an eclipse of the Moon. Note the Earth's shadow at the Moon is twice as wide as the Moon. Not to scale.

he knew the ratio of the sizes and distances of Sun and Moon (19), and the ratio of the sizes of the Sun and Moon compared to their distances from Earth (115). The rest is geometry, or trigonometry, whichever you prefer; the Alexandrians were adept at both. He computed the sizes and distances of the Sun and Moon relative to the size of the Earth. Again, you can do it yourself, although your version of figure 2-2 will be long and skinny indeed, more like a needle than a witch's hat. Get a big enough piece of paper so that the Earth, Moon, and Sun will be of a reasonable size—I use a roll of shelf paper in the classroom and roll it out across the floor—sketch it all out with the ratios above (remember to make the Earth's shadow at the Moon twice as wide as the Moon), and you'll see with Aristarchus that the universe suddenly gets VERY BIG compared to the Earth.

And since Eratosthenes had already measured the size of the Earth in stadia, Aristarchus now knew the actual sizes and distances of the Sun and Moon *in stadia.* No small achievement!

Certainly Aristarchus was not the first person to speculate on these matters, and it is difficult to ascertain just how much of his *Treatise on the Sizes and Distances of the Sun and Moon* is original.

But since we have no reason to believe otherwise, we should credit his priority. What came next was almost certainly original with Aristarchus. For this we have the testimony of his younger contemporary Archimedes. None of what I will now describe is mentioned in Aristarchus's book; we rely on the fragmentary testimony of others.

Like everyone else of the time, Alexandrian astronomers believed the Sun circled a stationary Earth that resides immobile at the center of the cosmos. But if the Sun is the larger object, reasoned Aristarchus—as much larger as is a melon to a grape—why not imagine that the smaller Earth annually circumnavigates a larger central Sun, rather than the other way around! And so Aristarchus, perhaps first among all human beings, set the Earth in motion.

His colleagues no doubt raised formidable objections to this radical idea. First of all, the hoary weight of tradition: "Everyone knows the Earth is the center of the cosmos." Second, we have no sense of flying through space. And finally, if the Earth moves in a circle about the Sun, which resides in the center of the cosmos, then the relative positions of stars on the outermost orb of the cosmos should seem to change during the course of the year as we look at them from different positions, an effect called parallax. (See figure 2-3.) But the relative positions of the stars is constant, exactly as if they were viewed from a fixed position at the center of the celestial sphere. Aristarchus's hypothesis of a moving Earth not only violates tradition and simple credulity, said his objectors, but also runs counter to observation.

Aristarchus rose to meet the objection from observation. The relative positions of the stars would not show any *perceptible* variation in relative positions, he said, if the starry shell that encloses the universe is vastly larger than even *the circular orbit of the Earth.*

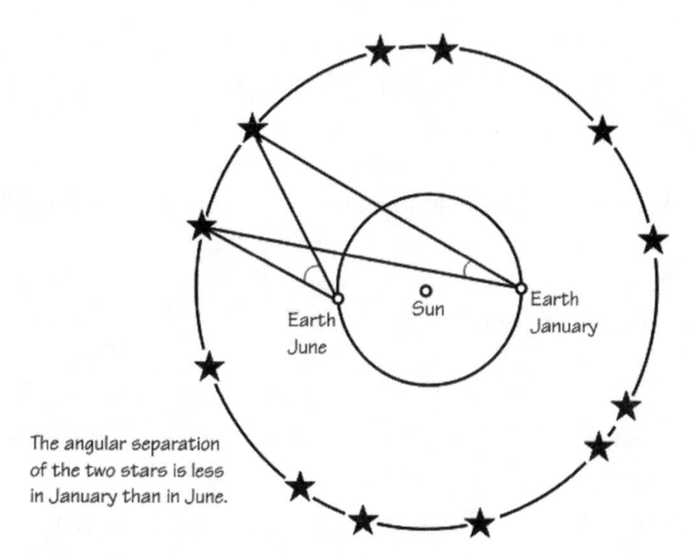

FIGURE 2-3. An example of parallax. Note that the apparent angular separation of two stars varies as the Earth orbits the Sun.

So commodious is the universe, he dared to assert, that the central Sun *along with the orbit of the circling Earth* are together *as a point* at the center of an enclosing sphere of stars of almost unimaginable size.

Not only this, but if the stars are so vastly far away, then they themselves—those cold points of light—must themselves be other Suns, reduced in brightness and apparent size by distance. And if the stars are other orbs of fire at vast distances, why should we assume that they revolve each day about a stationary Earth, rising in the east and setting in the west? Let the Earth spin each day on its own axis, and the appearances are saved—more simply, more economically.

And so Aristarchus gave the Earth a double motion: an annual revolution about the Sun and a daily rotation on its axis. All of this in a universe whose yawning dimensions bear no sensible relationship to the world of men and women. The cosmic egg is shattered; the Earth is sent spinning in a vast emptiness; the gods are removed to unimaginable distances beyond the stars. The stars are other suns, perhaps with other worlds. Possibly the universe is infinite! It is hard to imagine a mathematical deduction having more astounding implications. And remember, there is nothing here that you or I might not have done ourselves had we only the wit or genius to think of it.

Other astronomers of Aristarchus's time were no doubt aware of his bold proposals, but we have no evidence that any of them were prepared to agree with the revolutionary thinker. Aristarchus's determination of the sizes and distances of the Sun and Moon were one thing, possibly even correct, but putting the Earth into a double motion and removing the stars to unimaginable distances was simply a daring act of faith. The philosopher Cleanthes thought Aristarchus should be indicted on a charge of impiety; to imagine such a commodious universe was an insult to the gods, not to mention to our own firmly held conviction of divinely appointed centrality.

How unfortunate that the historical record is so scanty, that we know so little of Aristarchus and the genesis of his ideas. What sort of man is able to let his mind wander where no one's thoughts have gone before? He did not invent the method of mathematical and empirical speculation that we now call science, but he followed the method to its logical conclusion, even into heresy.

It turns out that Aristarchus was inexact in two of his crucial observations. He measured an angle of 87 degrees between the Sun and half-full Moon with whatever instrument he was able to muster. It is difficult to say when the Moon is *exactly* half full and even more difficult to precisely measure the angle between Sun and Moon at that time. The correct angle is 89.86 degrees. (You'll have a hard time drawing *that* triangle with your protractor.) It turns out that the Sun is not nineteen times farther away than the Moon; it is actually four hundred times more distant. Aristarchus also overestimated the apparent size of the Sun and Moon in the sky, and this too affected his calculation. The Sun's diameter is not six or seven times greater than that of the Earth; it is more than one hundred times bigger! *A million Earths would fit into the globe of the Sun!* Aristarchus's reasoning was impeccable, his mathematics flawless. It is perhaps just as well that his angle measurements were inexact; if he had discovered the true sizes and distances of the Sun and Moon, perhaps even *his* courage would have faltered in the face of the celestial abyss.

On June 22, 1633, Galileo Galilei was condemned by a tribunal of the Roman Catholic Church for teaching that the Earth is not the physical center of the universe but rather revolves about the Sun, as proposed almost two thousand years earlier by Aristarchus. Galileo had himself been convinced of the truth of Aristarchus's ideas by the Polish astronomer Nicolaus Copernicus, who showed that a heliocentric view of the cosmos provided a rather more elegant way to mathematically account for the motions of the planets. On his knees before assembled princes of the church, the seventy-year-old Galileo recanted his belief in the Earth's motion and renounced his life's work. In so doing, he escaped torture or perhaps even death at the stake, and won instead the lighter sentence of house arrest in Florence. An old story has it that after reciting the official recantation, Galileo whispered under his breath, "Eppur si muove" (Yet it moves). Whether or not he actually uttered the legendary words hardly matters; he surely must have thought them. He returned to Florence, frail and blind, and continued his experiments in physics. And the Earth went on revolving about the Sun.

How did things come to this sorry pass, where the greatest scientist of his time—and one of the greatest of all times—was forced to grovel before churchmen and deny what he believed to be true? Certainly, Galileo had reason to be afraid. Only thirty-three years earlier, in February of 1600, Giordano Bruno was burned at the stake in the Campo de' Fiori (Field of Flowers) in Rome for teaching, among other heresies, that the Sun was but one of an infinity of stars in a universe with a multitude of centers. Bruno was not a scientist in the mold of Aristarchus, Copernicus, or Galileo. He was a dreamer who let the scientific search for truth inform his dreams. His vision of an infinitude of possibly inhabited worlds was not based on any empirical evidence, only on an intuition of what was in keeping with Copernican cosmology: If the Earth is not the center of the cosmos, why single out our Sun as special? Why not assume that the stars that spangle the night sky are other suns, at vast distances from Earth, and centers of other planetary motions? Perhaps even Aristarchus had imagined as much.

Giordano Bruno was born in the Kingdom of Naples in 1548, only a few years after the death of Copernicus. At the age of twenty-four he was ordained a Dominican priest, although his curious and uninhibited mind had already earned the disapproval of his teachers. Within a few years of ordination he was accused of heresy—for the first of many times. The very idea of heresy meant nothing to Bruno; he claimed for himself (and for others) the *libertes philosophica,* "the right the philosophize, to dream," unfettered by authority or tradition. Poet, philosopher, and loose cannon, Bruno wandered across Europe—Italy, Geneva, France, England, Germany—stirring up a fuss wherever he went, offending both Catholics and Protestants, shaking up preconceptions, rattling complacencies, asking philosophers and shopkeepers alike to entertain a doubt or two: The universe and God, he said, might be bigger than we think. He was a modern in many qualities of mind: materialist, rationalist, a champion of free and skeptical inquiry. The universe is a unity, he believed; he made no distinction between matter and spirit, body and soul. He was energized—and exhilarated—by the majesty of the universe as it was being revealed by Tycho Brahe, Johannes Kepler, and other astronomer successors of Copernicus.

Within a decade of Bruno's execution, his dream of an infinity of worlds took substance. The vacant chair of mathematics at the University of Padua was offered to Galileo Galilei of Florence. In the winter of 1610 Galileo turned the world's first astronomical telescope to the night sky and saw things that would change the world forever: mountains on the Moon, spots on the Sun, moons of Jupiter, phases of Venus, and a myriad of tiny stars that twinkled beyond the limits of unaided human vision. He communicated these discoveries to the world in a little book called *The Starry Messenger,* in which he claimed, like Bruno, that the universe might be infinite and contain an infinitude of stars. Prudently, Galileo did not mention Bruno, although he surely knew of the radical philosopher and his unfortunate fate.

The Campo de' Fiori in Rome, the place where Bruno was burned, was a busy market square when I visited. A melancholy and somewhat sinister statue of the philosopher stands on a pedestal in the square, erected by secular humanists in the nineteenth century when the unification of Italy liberated the city from direct papal rule. Although the statue seemed dark and forbidding, and rather out of keeping with the personality I had imagined for Bruno, it stood among colorful stalls overbrimming with fruits, vegetables, and flowers, and *that* at least seemed appropriate for a man who celebrated so enthusiastically the beauty of creation.

In the late 1960s, early in my career as a science teacher, I made my way to Europe to visit the haunts of the scientists I was reading about in the history books. I wanted to do more than visit the shrines of science; I wanted to visit ideas too, in a more intimate way than merely reading secondhand accounts. During my yearlong stay at the Imperial College in London, studying with the eminent historians of science Marie Boas Hall and A. Rupert Hall, I used part of my time to apply the theories of each of four great astronomers to the motion of the planet Mars during that current year, 1968–69. I chose Mars because it is the planet most fully treated by the quartet of astronomers whose work I chose to study: Claudius Ptolemy, Nicolaus Copernicus, Tycho Brahe, and Johannes Kepler. Also, because Mars is a planet near to Earth, it lends itself to a compact graphical application of the astronomers' respective theories, and I have always loved playing with compass, protractor, and straightedge.

My first step was to obtain the positions of Mars along the zodiac during that current year. (The zodiac is the band of constellations around the celestial sphere that contains the motions of Sun, Moon, and planets.) My four astronomers, in developing and applying their theories, relied upon their own observations of the positions of celestial bodies or the observations of their predecessors. Since the time of the Alexandrian astronomers, the positions of celestial objects have been measured in degrees of longitude from the "first point of Aries," the vernal equinox, the Sun's place in the sky as it crosses the sky's (and Earth's) equator on the first day of spring. I did not have the opportunity or the instrument to make my own full suite of observations of the longitudes of Mars, so I took my data from published ephemerides of the planet. I did, however, measure a few Martian longitudes with a cheap, hand-held, plastic sextant—just to get a feel for how it was done.

Figure 2-4 shows the position of Mars in Earth's sky at fifty-day intervals from March 6, 1968, to December 17, 1969. Early in 1968 Mars was not far from the vernal equinox (and from the Sun) and moving, as usual, eastward against the background of the stars (as we look up into the equatorial sky from the northern hemisphere, east is to the left). It continued to move rather steadily eastward throughout the year, going halfway around the sky. Meanwhile the Sun overtook the planet and raced on ahead. Then, during the early months of 1969, now in the opposite part of the sky from the Sun, Mars slowed its eastward progress and for a few months moved *backward*, making a loop-the-loop in the sky, what astronomers both ancient and modern call retrograde motion. Mars quickly speeded up again. In all of this, the planet varied in brightness, becoming brightest of all when in retrograde and in opposition to the Sun (in the opposite part of the sky).

Other planets, too, have this curious habit of changing brightness and moving forward and backward against the backdrop of the fixed stars.

How to account for this behavior of the planets? The Greek astronomers, especially the Alexandrians—Hipparchus, Eudoxus, and Claudius Ptolemy—struggled valiantly to find a mathematical way to describe planetary motions. Like all Greeks, they were obsessed with the perfection of circles. Why, then, had the Creator not placed the planets on uniformly turning circles centered on the Earth? How much more elegant the cosmos would be, they must surely have thought, if the planets moved *steadily*

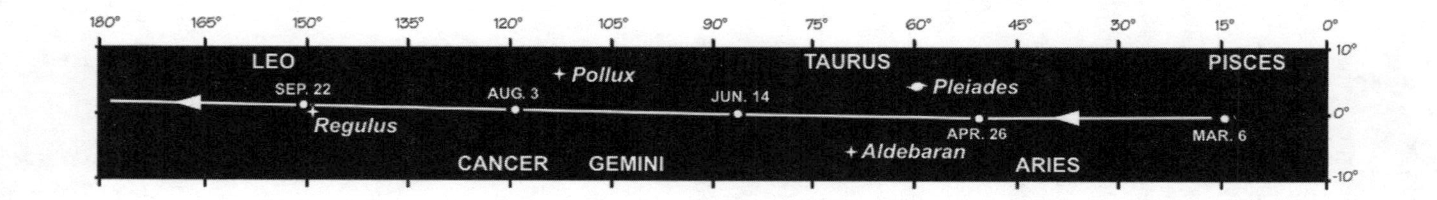

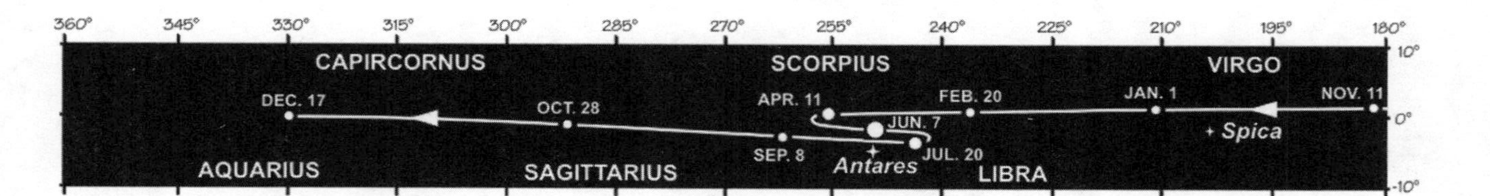

FIGURE 2-4. The apparent motion of Mars along the zodiac in 1968–69, at fifty-day intervals. The Sun (not shown) is in the same part of the sky at the extremes of this interval, and in the opposite part of the sky as Mars does its loop-the-loop.

across the sky without all this speeding up and slowing down, loop-the-loops, and changing brightness. Why wasn't the cosmos a neat arrangement of nested concentric spheres, like Russian dolls, each sphere carrying a celestial body, all turning uniformly about the fixed central Earth? All this loop-the-looping in the sky; what could the Creator have been thinking?

The Alexandrian astronomers were committed to uniformly turning circles by both aesthetic taste and the limitations of their mathematical tools. They were also committed—with the exception of Aristarchus—to the *centrality* of the Earth. So they sought some combination of uniformly turning circles that would carry the planets along courses that matched what they saw in the sky. To this end, they invented three useful devices.

The first and most important was the *epicycle.* The planet moved on a uniformly turning circle (the epicycle) whose center was fixed to the rim of a larger uniformly turning circle (the deferent) whose center was the Earth. With appropriate choices of rates of turning, this device neatly gives the planet a loop-the-loop motion and accounts for variations in brightness by allowing the distance of the planet from Earth to vary.

But still the match of theory and observation was not perfect. So the Alexandrian astronomers found it helpful to give the planet's deferent a center (called the *eccentric*) that was not quite the Earth and to let the deferent's motion be uniform about yet another point (the *equant*) offset still farther from the Earth by an equal amount along the same line. All of this—the sizes of the deferent and epicycle, their rates of turning, the distances and direction of the eccentric and equant—were adjusted until this strange mathematical apparatus yielded positions that "saved the appearances"; that is, that matched the observed positions of the planet in the sky.

The person who brought all of this to a high state of perfection was Claudius Ptolemy, of whom we know little except that he lived in Alexandria in the second century of the Christian era. His name suggests a Greek-Egyptian ancestry (Ptolemy) and perhaps Roman citizenship (Claudius). This is the same Claudius Ptolemy who compiled the atlas that represented the culmination of Alexandrian geography; he was clearly a man of wide interests and accomplishment. It is a miracle that his book on astronomy, *The Almagest,* has come down to us, through more than a millennium of copying by hand before it finally found its way into print at the time of the European Renaissance. The title comes from the Arabic translation (*al-majisti*) of an earlier Greek title, *The Greatest Compilation,* and simply means "The Greatest." The work consists of thirteen books, of which the final five discuss Ptolemy's planetary theory. I present here (figure 2-5) my original drawing of Ptolemy's theory as I completed it more than three decades ago with straightedge, compass, and protractor. I can't promise that I got everything exactly right, but a comparison of figures 2-4 and 2-5 shows how satisfactorily the theory describes the motion of Mars through the sky in 1968–69—the changing speed, the loop-the-loop, the changing brightness.

What does it mean? Could the Creator of the cosmos have had such curious mathematical devices in mind when he set the planets in their courses? There is a shattering disjunction between the *simplicity* and *elegance* we associate with great creative art—and which Plato and Pythagoras assumed of the Creator—and Ptolemy's crazy clockwork of deferents, epicycles, eccentrics, and equants. What we *do* see in Ptolemy, however, is an almost fanatical fidelity to quantitative observational data. What was most important to the Alexandrian astronomers was not a system of the world such as they would have assumed of a Creator with Greek sensitivities, but a system of the world that agrees with quantitative

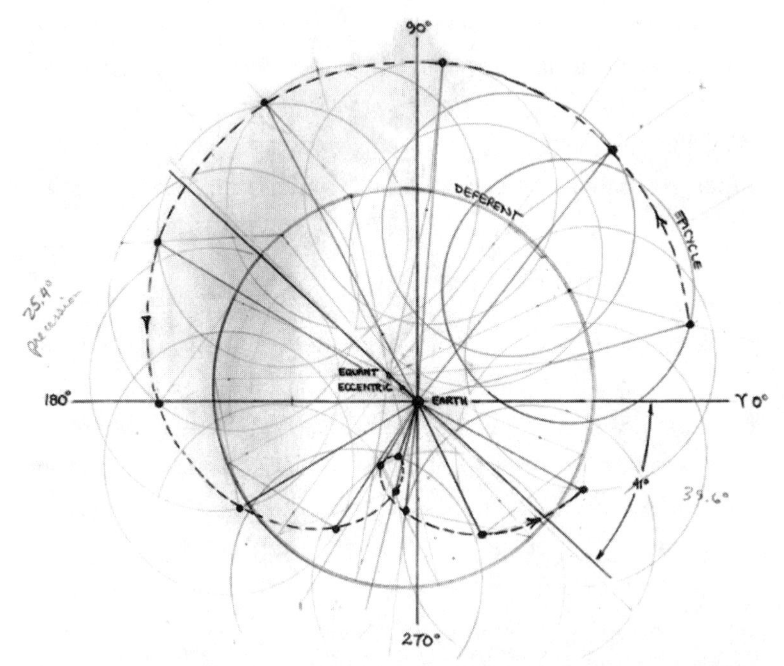

FIGURE 2-5. Claudius Ptolemy's theory applied to Mars, 1968–69, incorporating deferent, epicycle, eccentric, and equant.

observation. In other words, the Alexandrian touchstone of truth was not *what we would like to be true,* but what our senses affirm to be true—a lesson many of us still have trouble embracing. For its ability to describe and *predict exactly* the motions of celestial bodies, Ptolemy's system of the world stood unsurpassed for 1,400 years.

During all of that time it must be said that Alexandrian astronomical theories had very little influence on how the great majority of people *thought* about the world. The Romans were a practical people with little taste for mathematical abstractions, and with the rise of Christianity empirical science took a backseat to theological speculation. The cosmos depicted in medieval European art and philosophy is a cozy human-centered nest of concentric circles, presided over by a God who created the whole shebang for one reason only: as a stage for the human drama of sin and salvation. In medieval Christian cosmology angels trump angles, and a nostalgic longing for the parent's secure embrace is considered more important than the actual motions of planets in the sky.

Nicolaus Copernicus (1473–1543) was born in Poland of a well-to-do family and studied in Italy as a young man before settling as a canon of the Roman Catholic Church at Warmia, where his uncle was bishop. From what we can tell, Copernicus was adept at medicine, law, mathematics, and astronomy. Astronomy seems to have been his greatest love, and he was apparently deeply unsatisfied by the Ptolemaic system of the world. Ptolemy's theories might indeed "save the appearances," but Copernicus felt there must be a more elegant way to do it. He was drawn to the ideas of Aristarchus, who had recognized many centuries earlier that our view of the world might be simplified by letting the Earth turn once each day on its axis, rather than having everything else, including the myriad stars, turn about the Earth. Aristarchus also conceived of an Earth that raced in orbit around the Sun. But Aristarchus was ahead of his time, or rather, ahead of common sense. If the Earth turns each day on its axis, we must be flying along at roughly a thousand miles per hour, and nothing in Greek or medieval European physics could explain why it is we don't notice such a breathtaking flight. There were good scientific reasons to stick with an immobile Earth.

Copernicus could provide no explanation for why we don't feel our motion on a spinning Earth (for that, Galileo and Newton would invent a new kind of physics), but he thought the theoretical advantages of having the Earth turn on its axis outweighed the affront to common sense. As for why the apparent positions of the stars do not change as the Earth revolves around the Sun, well, he had a explanation for that too, the same as that of Aristarchus: The stars are so far away compared to the size of the Earth's orbit that their parallax is imperceptible. So, like his Alexandrian predecessor, Copernicus wrenched the Earth from the center of the cosmos and sent it whirling through a space that was vast beyond comprehension.

Copernicus's great book, *On the Revolution of the Heavenly Spheres,* was published in 1543, the author receiving his copy on his deathbed. He had waited a long time to publish, and did so finally only at the urging of his young Protestant friend Georg Joachim Rheticus. Even then, Copernicus, a Catholic churchman, knew his theory would be considered heretical, undoubtedly by Catholics and Protestants alike. It was, however, a time of displacement from centers. Brave mariners had discovered previously unknown civilizations on new continents. Martin Luther had nailed his ninety-five theses to the door of the Wittenberg church in 1517, and Rome was no longer the only center of Christian doctrine. The invention of the printing press had placed sacred scriptures in the hands of ordinary folks who could read for themselves the word of God. It was also a time of extraordinary creativity and attention to the natural world; the artist Albrecht Dürer, for example, could sit down before a patch of meadow and render it with such precision that today we can identify in his drawing every species of plant. Copernicus lived in a revolutionary time and added his own jolting innovation to the turmoil.

In figure 2-6 you can see my application of Copernicus's theory to the motion of Mars in 1968–69. The Earth's circular orbit takes the place of Ptolemy's epicycle. The orbit of Mars is eccentric to

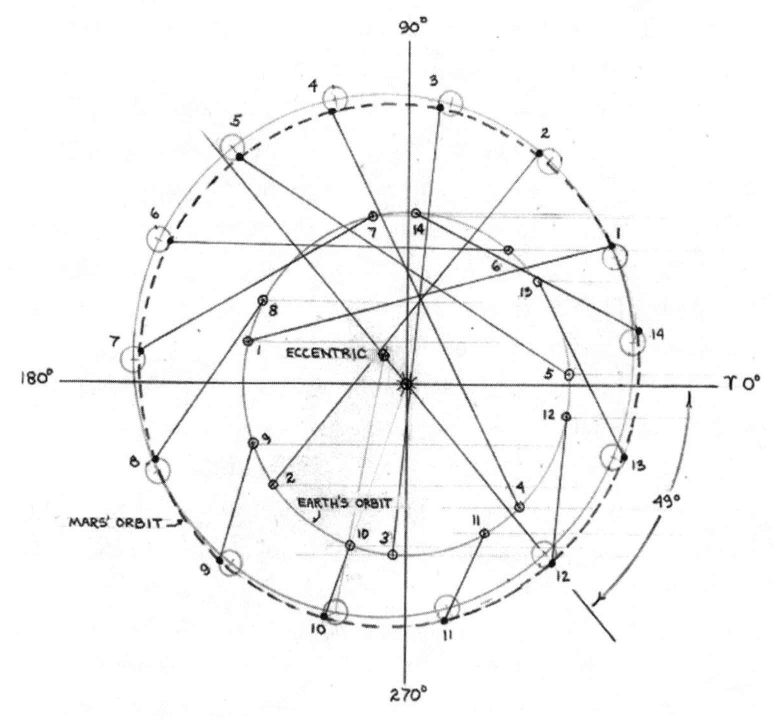

FIGURE 2-6. Nicolaus Copernicus's theory applied to Mars, 1968–69.

the central Sun, so here is another echo of Ptolemy's theory. And that little epicycle you see rotating on the big circle of Mars's orbit in effect replaces Ptolemy's equant. If you follow the lines of sight from Earth to Mars in my drawing, you will see that Copernicus's system reproduces the observed motion of Mars in 1968–69 with a fidelity equal to that of Ptolemy. The apparent retrograde motion of Mars occurs as the Earth approaches and overtakes the red planet in their respective orbits.

So where is the gain in simplicity? Doesn't Copernicus's theory for Mars employ the same number of artificial devices as does Ptolemy's? Yes. But here is the advantage: In Copernicus's system, the Earth's orbit around the Sun replaces the epicycles of *all* of the planets, and thus the whole system approaches more closely a nest of concentric circles. Having worked through both theories, I can confirm that they are about equally difficult to handle geometrically, at least when applied to a single planet. But I am confident that for Copernicus the heliocentric version *just looked right,* more in keeping with a rational Creator's desire for simplicity. He wrote: "At the middle of all things lies the Sun. As the location of this luminary in the cosmos, that most beautiful temple, would there be any other place or any better place than the center, from which it can light up everything at the same time? Hence the Sun is not inappropriately called by some the lamp of the universe, by others its mind, and by others its ruler."

So the Earth orbits a central fire, like all the other planets. But wait! Not quite *all.* In Copernicus's system the Moon still moves about the Earth. So the Earth retains a measure of centrality as the center of the Moon's orbit, a *specialness* that sets it apart from the other planets. The Copernican system has *two* centers, and this must surely have troubled Copernicus and those who took up his cause. Remember, until Galileo turned his telescope to the sky no one knew that other planets had moons.

Certainly, the world did not rush to embrace the new Copernican cosmology. For one thing, the man who supervised the printing of the book, the Lutheran theologian Andreas Osiander, inserted a preface ostensibly from the hand of Copernicus stating that the new system of ascribing a double motion to the Earth—rotation on its axis and revolution about the Sun—was not meant to be taken literally; it was merely a simpler way to calculate the positions of heavenly bodies. Presumably, according to the preface, Copernicus continued to believe that the Earth was the fixed center of the cosmos, even as he used a heliocentric fiction to make predictions. It would be fifty years before Osiander's preface was exposed as a distortion of Copernicus's own views. But perhaps Osiander knew what he was doing. So prefaced, the book escaped immediate condemnation by ecclesiastical authorities and made its way relatively unscathed into the world.

One person who recognized the advantages of the Copernican system was the Danish astronomer Tycho Brahe (1546–1601), a brash nobleman with a silver nose (he lost his fleshy appendage in a duel) and an island estate in the strait between Denmark and Sweden. His house might fairly be called the first modern observatory. Although the telescope would not be invented until a few years after his death, Tycho assembled the best nontelescopic instruments money could buy and set about measuring the positions of celestial objects with unprecedented accuracy. He also tinkered with his own mathematical system of the world, a rather cumbersome amalgam of Ptolemy and Copernicus, with the Sun going around a central Earth and the planets circling the Sun. Tycho was trying to have his Copernican cake and eat it too; he was unwilling to make the biggest break of all: displacing humankind from the center of the cosmos. Figure 2-7 shows Tycho's theory applied to Mars in 1968–69, and you can see that it does the job. As far as I know, however, no one but Tycho took his system seriously.

The person who took Copernicus *most* seriously was Johannes Kepler (1571–1630), a man consumed with a desire to know the true nature of the world as revealed by the senses, and yet as committed as any Greek to the idea that the Creator worked from a simple mathematical plan. In his earliest published work (1596) he proposed to forget the various whirligigs of turning circles

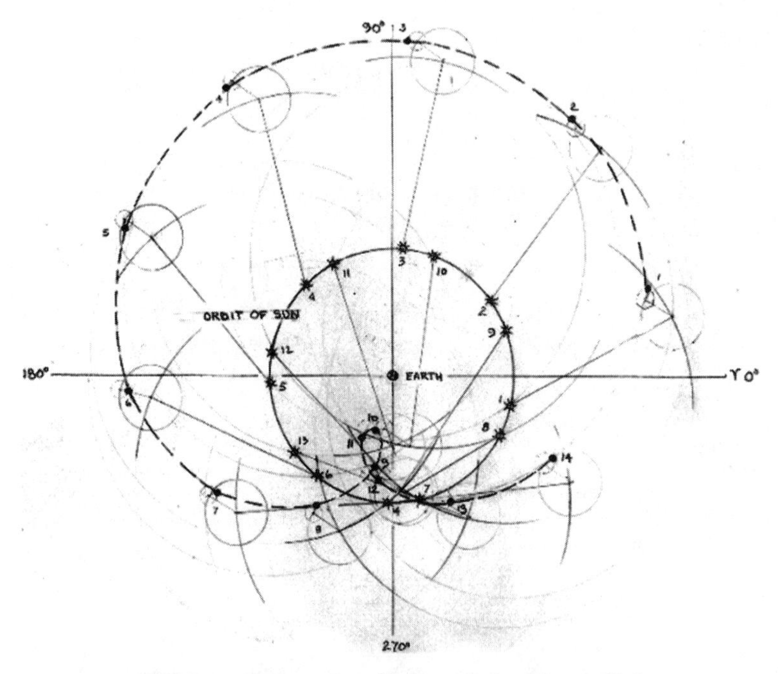

FIGURE 2-7. Tycho Brahe's theory applied to Mars, 1968–69.

employed by his predecessors and instead determine from direct observation what were the actual paths of planets in space. Kepler knew that Tycho had the data he needed, the best and most precise observations currently available for planetary motions. He managed to have himself appointed as Tycho's assistant. (Tycho was then working in Prague as imperial mathematician.) On Tycho's death, Kepler inherited the Danish astronomer's observations for Mars.

What followed was one of the epic bouts of calculation in the history of science, what Kepler called his "war with Mars," thousands of pages of tedious arithmetic as he attempted to extract from Tycho's observations the true path of the planet, all while afflicted with illness, weak eyesight, religious persecution, and interminable financial and family problems, including an almost endless struggle to keep his quarrelsome mother from being burned as a witch. He published his results in *Astronomia Nova* (The New Astronomy) in 1609. In 1968–69 no English translation of Kepler's original text was available, but I found in a North London bookshop a volume that gave me the information I needed, Robert Small's *An Account of the Astronomical Discoveries of Kepler,* published in English in 1804. (It was the purchase of Small's volume that inspired my desire to work my way through the systems of Ptolemy, Copernicus, Tycho, and Kepler.) Figure 2-8 shows the result. No circles turning on circles. No epicycles or equants. The orbit of Mars (like that of Earth and the other planets) is a pure geometric ellipse with the Sun as a focus. The ellipse is almost

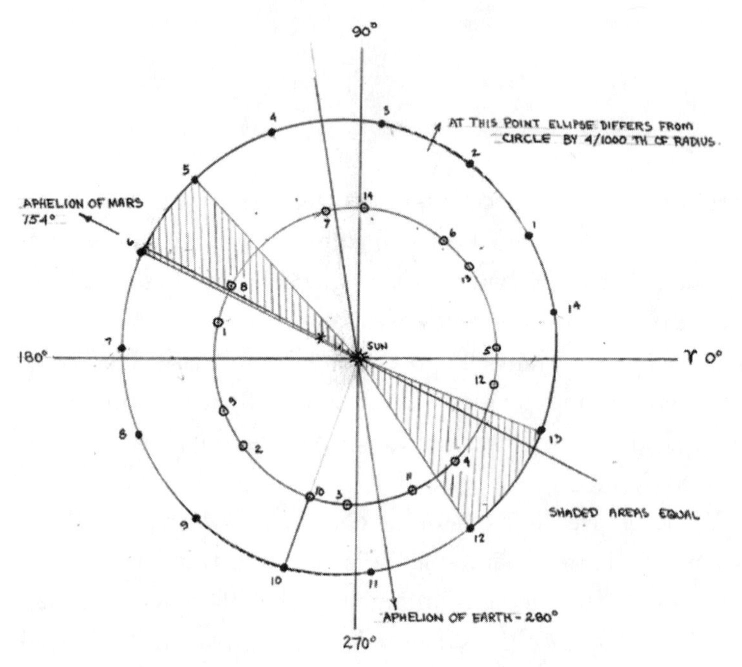

FIGURE 2-8. Johannes Kepler's theory applied to Mars, 1968–69.

imperceptibly different from a circle, not at all the elongated egg-shaped curve one usually sees in textbook illustrations of Kepler's theory. The planet does not move uniformly on its ellipse; it moves faster when close to the Sun and slower when farther away. Kepler found a law, however, that describes the planet's varying speed: In equal intervals of time Mars sweeps out equal areas about the Sun. (See the shaded areas on my diagram for two fifty-day intervals.) And further, Kepler discovered a simple mathematical rule to relate the periods of the planets (the times for a complete orbit) to their mean distance from the Sun: The squares of the periods are proportional to the cubes of the distances.

These, then, are Kepler's three laws of planetary motion:

1. The planets, including the Earth, move in elliptical paths with the Sun as a focus.

2. They sweep out equal areas about the Sun in equal times, moving faster when close to the Sun, slower when farther away.

3. The squares of their periods are proportional to the cubes of their mean distances from the Sun.

So God, it seems, is a mathematician after all; he is just not as attached to circles as the Alexandrians and Copernicus had imagined. Nor was it the Creator's intent, presumably, to place humankind at the center of his design.

Each of Kepler's laws of planetary motion makes reference to the Sun. His system is even more profoundly heliocentric than that of Copernicus. Kepler was quick to guess that the Sun might exert a controlling influence on all of the planets, perhaps in some way communicating its own spinning motion to them through an intervening medium. While the astronomical systems of Ptolemy, Copernicus, and Tycho were purely mathematical, Kepler's universe begins to take on *physical* characteristics; the Sun is not just a center but a causal agent. What that causal agency might be remained for Isaac Newton to discover.

This is perhaps a good time to pause and reflect on the old question: Do scientists invent or discover? Are the respective theories of Ptolemy, Copernicus, Tycho, and Kepler mere human artifacts, ways of "saving the appearances" as Osiander suggested, or do they capture some aspect of reality? After all, Ptolemy affirmed a central Earth with the same fervor that Kepler later claimed a central Sun. If one generation's truth is the next generation's folly, who is to say that truth can ever be the product of scientific inquiry?

First, no one, least of all scientists, claims that science yields truth—at least not Truth with a capital *T*. Science attempts to describe/explain the evidence of the senses in the most elegant and economical way possible, without invoking other than natural causes. The astronomical theories we have just considered each concisely encapsulates observations of celestial motions, with more or less equal fidelity. Moreover, each theory enables astronomers to reliably predict future celestial phenomena; that is, the theories confer practical benefits, such as refining calendars, setting times of religious festivals and rites, and predicting eclipses. The last of these theories, that of Kepler, remains today the most concise and elegant way we have to describe celestial motions.

Kepler's account of planetary orbits would later be incorporated into Newton's physics and lead to the discovery of new planets: Uranus and Neptune. It also describes the orbits of every craft we have hurled into space. Artificial Earth satellites move in elliptical orbits according to Kepler's laws of planetary motion, as do the spaceships we send to other planets. Kepler's theory may not be True, but it is certainly more reliably predictive and practically useful than any other way we have yet invented for describing celestial phenomena. That is, the theory may not be True, but it is certainly truer than alternatives, and so far we have no reason to doubt its veracity, and every reason to affirm it.

There is no such thing as the infallible Truth-generating "scientific method" often attributed to Galileo's contemporary Francis Bacon—pose a hypothesis, do a experiment, refine the hypothesis, et cetera—all very impersonal and guaranteed to lead inevitably to the raw reality of nature itself. As the biologist Stephen Jay Gould pointed out, Bacon himself understood that science is (in Gould's words) "a quintessential human activity, inevitably emerging from the guts of our mental habits and social practices, and inexorably intertwined with foibles of human nature and contingencies of human history." The theories of Ptolemy or Kepler can only be understood within the context of their creators' personalities, learning and experience, cultures, times. Understanding empirical data means searching for suitable metaphors, analogies, and patterns of meaning, and what we find suitable is culturally (perhaps even genetically) determined to a greater extent than the ultrarealists are willing to admit.

Are theories, then, entirely arbitrary cultural constructs, as the ultrarelativists assert? No. In Bacon's words, scientific understanding "is extracted . . . not only out of the secret closets of the mind, but out of the very entrails of Nature." An acceptable scientific idea must be consistent with ever more finely contrived observations of nature. The theories of Ptolemy, Copernicus, and Kepler are severely constrained by celestial observations, and it is this that gives us confidence that each of the theories possesses a measure of truth. Contemporaries of Copernicus and Kepler believed that comets were divinely ordained signs and portents. When it was discovered that comets obey precisely Kepler's laws of planetary motion, the idea of portents went out the window. Only the most hard-nosed relativists would deny that Kepler's theory of comets is in some sense truer than the theory of portents.

Scientists hold their theories against the refining fire of experience. Paradoxically, the constraints of data might actually promote scientific creativity. The poet Robert Frost famously said that writing unrhymed verse is like playing tennis with the net down. He understood that artistic creativity is enhanced, rather than frustrated, by working within certain formal constraints. The best poets all impose complex structural restrictions upon their work: rhyme schemes, sound patterns, syllabication, and so on. Aspiring young poets sometimes believe that by merely emptying the closets of their minds willy-nilly onto paper they have created poetry. It's not poetry that they have created, and it's not particularly creative. Poetry is a very special use of language in resonant tension with the world. In the same way, scientific creativity is sharpened, not dulled, by rubbing against the whetstone of reality. Empirical observation constrains the scientist's creativity but does not force inventiveness along inevitable tracks. It is for the scientist as it was for the singer in a poem by Wallace Stevens: "Even if what she sang was what she heard . . . there never was a world for her / Except the one she sang, and singing made."

What we seek in science is not Truth but knowledge more reliably true than any alternative.

3

ANTIQUITY OF THE EARTH

Between Peacehaven and the town of Lewes, my path takes me across chalky agricultural uplands. It is one of the glories of the English countryside to be crisscrossed with public footpaths; I will easily make my way from town to town without ever straying far from the line of zero longitude. It is another glory of the English countryside that strict zoning has prevented urban and suburban sprawl, so that the landscape between towns remains relatively unspoiled. On my trek from Peacehaven to Lewes I have not put foot on asphalt, except for a brief swing through the fairy-tale village of Telscombe, tucked sleepily in a dale where it has reposed since the Middle Ages.

As I approach the end of my day's walk, I look down from a crumbly white chalk precipice onto Lewes, an ancient market town on the River Ouse. The path descends sharply, traverses a long bare ridge, jumps the A27 bypass highway, and suddenly I'm on the main thoroughfare of the town and quaffing a pint of beer in the Meridian Pub. Just outside the door of the pub a plaque set into the pavement marks the Greenwich meridian.

My libation finished, a few dozen more paces along the high street brings me to a handsome old house with another plaque, this one designating the former residence of Dr. Gideon Algernon

Mantell, eminent surgeon and self-made geologist, former resident of Brighton and Lewes, and discoverer of dinosaurs.

The traditional story has it that Mantell's wife, Mary Ann, found the first relic of extinct giant terrestrial reptiles—a fossilized tooth—on a family outing to Cuckfield, a village some ten miles north and west of Lewes. Supposedly, she noticed a curious stone in a wayside wall and brought it to her husband, who recognized immediately that he held something new and extraordinary in his hand, a tooth like no other he had seen before. This story, however, has no evidential basis. The tooth that came into Mantell's hand was probably passed to him in late 1820 by a quarryman at Cuckfield, who would have known of the doctor's long-standing interest in fossils. At that time Mantell was putting the finishing touches on his book, *The Fossils of the South Downs,* an account of the aquatic creatures that left their impressions in the chalky strata around Lewes. (The former age in which the chalk was deposited in a marine environment is called Cretaceous, from the Greek word for "chalk.") Among the ancient animal remains preserved in Mantell's collection were starfish, corals, sea urchins, and seashells of many kinds, plus extinct creatures such as crinoids and ammonites. Some vertebrate specimens reposed in the chalk: shark teeth and the skeletons of fishes. The Cuckfield tooth was very different from the fossils of the chalk. First, it came from strata underlying the chalk and therefore was presumably older. Second, whatever creature possessed the tooth was larger than any known animal that lived in Cretaceous seas. Third, Mantell was of the opinion—correctly, it turned out—that the well-worn tooth he held in his hand belonged to a land-dwelling, herbivorous reptile.

Where there is a tooth there must be bones, and the Cuckfield quarry soon yielded them up too, enormous things, as large as the bones of the largest animals existing today, interbedded with an assemblage of fossils different from any that Mantell had collected from the chalk. Ferns and other land plants. Turtles. Birds. Crocodiles. The teeth and scales of fishes. Some marine fossils but mostly creatures of the shore. Clearly these were relics of a time when southeastern England was near the edge of a shallow sea, alternately dry and inundated. Among the animals that resided on that strange shore was a creature that soon came to be called iguanodon, meaning "iguana tooth," because of the resemblance of the Cuckfield tooth and bones to present-day—although smaller—iguana lizards.

In 1831 Mantell published a paper in the *Edinburgh New Philosophical Journal* that begins: "Among the numerous interesting facts which the researches of modern geologists have brought to light, there is none more extraordinary and imposing than the discovery that there was a period when *the earth was peopled by oviparous quadrupeds of a most appalling magnitude,* and that reptiles were the *Lords of the Creation,* before the existence of the human race!" Oviparous quadrupeds. Appalling magnitude. Egg-laying, four-legged giants. Mantell's announcement was breathless and received with enthusiasm. The decade previous to his Cuckfield finds had been an extraordinary period in the history of geologic science. With surprising alacrity the sedimentary strata of southern England had yielded the fossilized remains of a strange family of animals now vanished from the Earth, aquatic and terrestrial, large and small, carnivorous and herbivorous. Mantell was but one of many eager explorers who teased the bones of iguanodon and other reptilian creatures out of the strata and into the popular imagination. The eminent geologist and paleontologist

Richard Owen proposed the collective name *dinosauria* for this extinct race of beasts, meaning "fearfully great lizards."

Among the fossil hunters who opened our eyes to the Earth's reptilian past, none is more justly famed than Mary Anning of Lyme Regis in Dorset. Her hunting ground was some 130 miles west of the prime meridian, among Jurassic strata tens of millions of years older than those in which Mantell found iguanodon, but many of Anning's fossils can be seen today in the Natural History Museum in London—very much along the line of my walk.

She came from humble origins and was not adverse to scrambling among the crumbling strata of Dorset's coast in voluminous skirts. At the age of twelve, she found what turned out to be the world's first ichthyosaur (fish-lizard), an extinct reptile so perfectly adapted for sea life that in some respects it resembles a fish, with fore and hind legs modified to function as fins. This is not to say that no one had previously observed ichthyosaur fossils (quarrymen had been turning out these mysterious creatures from the rocks for centuries), but not until Anning's specimen was examined did geologists recognize that what they were looking at was an animal unlike any that exist on the planet today. (See figure 3-1.) Soon more and better specimens came tumbling from the cliffs. The stony ichthyosaur impressions, when entire, were typically twelve or fifteen feet long, with long snouts and gaping eye sockets, and their prevalence in the rocks of Lyme Regis suggests that the animals were quite common in Jurassic seas. During her brilliant career as a fossilist, Anning found many kinds of ichthyosaurs, as well as plesiosaurs (near-reptile, another large marine vertebrate), and Britain's first pterosaur (a flying reptile).

FIGURE 3-1. An ichthyosaur from the collection of the Natural History Museum, London.

The Lyme Regis of Anning's time was a popular summer tourist destination—as it is today—and she was able to make a modest living selling common fossils as souvenirs. Her more important finds made their way to the eminent savants of London's Geological Society, founded in 1807 to facilitate interest in the emerging science of geology. Although Anning was as knowledgeable about these fossil creatures as many of her university-educated scientific colleagues, it was her fate to be female, of low birth, and of a dissenting faith, all of which denied her the kind of acclaim that accrued to Gideon Mantell. The gentlemen geologists of London recognized Anning's talent, but they did not often give her due credit when it came time to publish descriptions of her fossils.

Several of Anning's finest ichthyosaur and plesiosaur specimens reside today in the Natural History Museum in London, including her largest ichthyosaur, in what I like to call "the Hall of the Flat Beasts," a long gallery whose walls are adorned from floor to ceiling with fossilized marine reptiles splendidly displayed, including a few plaster replicas of important specimens in foreign museums. (See figure 3-2.) No matter how rotund these creatures were in life, when they fell dead to the sea floor and became buried in sediment their skeletons collapsed or were pressed nearly flat, so that when they are revealed in the strata they have an intaglio appearance, like carvings by Renaissance masters. The gallery has not changed since I first visited the museum in 1968. Most of the museum's other exhibit spaces have been jazzed up and made more user-friendly, but it would be hard to imagine how the Hall of the Flat Beasts could be improved; I hope the curators have the good sense to leave it alone, since it happily conveys the passion for collecting that characterized early nineteenth-century England.

One ichthyosaur specimen in the gallery (from Germany) contains six unborn young inside its body, and another has three

FIGURE 3-2. Marine reptile display, Natural History Museum, London.

unborn young with the almost perfect impression of a fourth being born tail first just as the mother died. A Lyme Regis ichthyosaur has bits of another ichthyosaur between its teeth, part of the creature's last meal. To move along the gallery from specimen to specimen is like being taken back 200 million years to vanished seas teeming with monsters—eating, being eaten, mating, bearing young. When I visited the gallery in 2003 during my meridian walk, an adjoining room was offering a special exhibit on that favorite of all dinosaurs, *Tyrannosaurus rex,* complete with an animated life-size model. As I sat on a bench in the Hall of the Flat Beasts among the seafaring monsters of Jurassic seas, every now and then from the other room came the disconcerting roar of carnivorous, land-trodding *Tyrannosaurus rex,* calling from deep geologic time. Gideon Mantell, Mary Anning, and their contemporaries did for time what the Alexandrians did for space: They extended it far beyond the bounds of here and now.

What did the early nineteenth-century fossilists make of the strange creatures in the rocks? Mantell and Anning lived in a place where one concept of time was then being challenged and a new concept of time was being born. Most people in England in the early nineteenth century held to a literal biblical view of Earth's history, which was coeval with human history except for six preparatory days as the Creator readied the stage for the human drama. Unlike France, which had cast off much of the intellectual influence of the church during its revolution, England still languished under the pervasive influence of the established Anglican faith. Even the great William Buckland (1784–1856), the first professor of geology at Oxford University and the nation's dominant authority in all things geologic, held to the view that the rock strata and their enclosed fossils could somehow be reconciled with the biblical account of creation and the flood of Noah. Buckland's friend and fellow fossilist Reverend William Conybeare also struggled to find a way to wedge the fossils into Genesis.

France's most prominent expert on fossils was Georges Cuvier (1769–1832), a man of commanding intellect, who thought the Bible irrelevant when it came to interpreting the creatures in the rocks. Cuvier was an anatomist by training and brilliantly qualified to understand the relationship between fossilized skeletons and animals that exist today. It was clear to him that many creatures that roamed the land or swam the seas in past times had become extinct, and that many catastrophic upheavals must have punctuated the long history of life. What these so-called revolutions might have been he left rather vague, but it was clear to Cuvier that the record of the sedimentary rocks confirmed an antiquity for the Earth vastly greater than the mere thousands of years implied by Genesis.

In Britain the antiquity of the Earth was most forcefully espoused by the Scottish gentleman farmer James Hutton (1726–97), who took a decidedly different tack than did Cuvier. Hutton's *Theory of the Earth,* published in 1788, is a prolix, almost unreadable book, but it contains one great idea that would lift geology out of the confining strictures of biblical exegesis. *Past geologic phenomena must be accounted for by the same physical processes we see at work today,* said Hutton. Where we find strata of sedimentary rocks, we must suppose they were deposited grain by grain by the same almost imperceptibly slow forces of erosion, transport, and deposition that even today remove the bulk of mountains and build up beds of mud or sand in inland lowlands or along coasts.

Where we find fossils in the strata, we must suppose they found their way into the rock in the same way that seashells and the bones of animals are sometimes buried in sediments today.

Hutton's view of Earth history is not one of recurrent catastrophes, à la Cuvier, but of continuous gradual transformation—mountains rising and falling by infinitesimal degree. No acts of divine intervention are necessary to explain the record of the rocks, said Hutton, only the natural forces of change that we watch at work today—a hypothesis that came to be called *uniformitarianism.* But if mountains are lifted and removed millimeter by millimeter, if thick beds of sediment are deposited grain by grain, then the six thousand years or so allotted to the Earth by biblical exegetes are manifestly insufficient. The Earth must be very ancient indeed, *and geologic time is something vastly different from human time.* How many years were required to explain geologic phenomena Hutton was not prepared to say, but certainly millions, perhaps more.

Hutton's prose may have been opaque, but the evidence of the rocks was compelling. There is a singular moment in Hutton's story that I have often taken to represent the birth of geologic time. Not that Hutton was the first to guess at the great antiquity of the Earth; certain Greek thinkers, among others, conceived of time without beginning or end. But not until Hutton did empirical observation at last convincingly confirm the apparently limitless duration of geologic time. The moment I am thinking of occurred in the spring of 1785. Hutton invited his friend John Playfair and a young man named James Hall to accompany him on a boat excursion along the Berwickshire coast of Scotland. From his observations of rocks on land, Hutton knew what he was looking for: a place along the coastal cliffs where two epochs of geologic history were simultaneously revealed. He found his

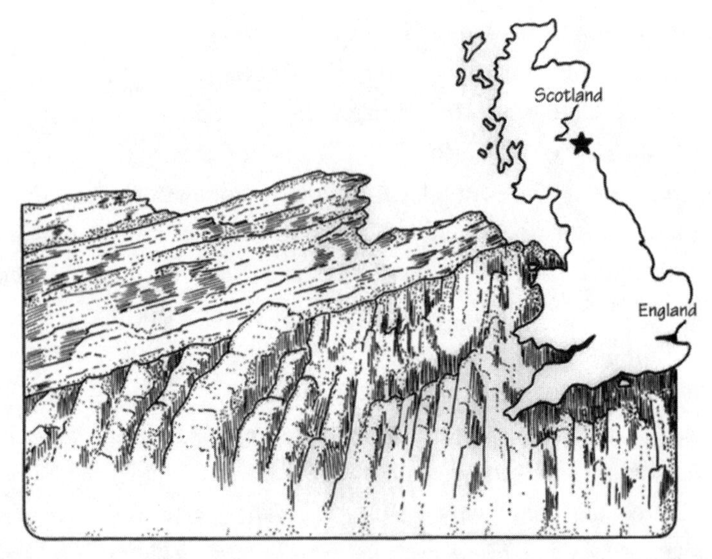

FIGURE 3-3. Unconformity of ancient sedimentary rocks, Siccar Point, Scotland.

sought-for disjunction of strata at a place called Siccar Point. (See figure 3-3.) Here were what appeared to be very ancient almost vertical strata, truncated by erosion and overlain with nearly horizontal strata of younger rock—what we would today call an unconformity—two episodes of deposition separated by a period of uplift and erosion, the surface of the Earth subsiding, rising, and falling again, tilted, folded, and eroded, all in slow motion, degree by infinitesimal degree. John Playfair responded enthusiastically to Hutton's exposition:

> On us who saw these phenomena for the first time, the impression made will not easily be forgotten. The palpable evidence presented to us, of one of the most extraordinary and important facts in the natural history of the earth, gave a reality and substance to those theoretical speculations,

> which, however probable, had never till now been directly authenticated by the testimony of the senses. We often said to ourselves, What clearer evidence could we have had of the different formation of these rocks, and of the long interval which separated their formation, had we actually seen them emerging from the bosom of the deep? We felt ourselves necessarily carried back to the time when the [vertical strata of] schistus on which we stood was yet at the bottom of the sea, and when the sandstone before us was only beginning to be deposited, in the shape of sand or mud, from the waters of a superincumbent ocean. An epocha still more remote presented itself, when even the most ancient of these rocks, instead of standing upright in vertical beds, lay in horizontal planes at the bottom of the sea, and was not yet disturbed by that immeasurable force which has burst asunder the solid pavement of the globe. Revolutions still more remote appeared in the distance of this extraordinary perspective. The mind seemed to grow giddy by looking so far into the abyss of time.

Here is an episode in human intellectual history that is worth savoring—a brief, seemingly inconsequential spring day on the coast of Berwickshire, three friends messing about in a boat and clambering on the rocky shore. It is 1785. Great events are afoot in the world. In Britain the Industrial Revolution is in full swing; soon the ten-thousand-year-long Age of Agriculture will give way to the Age of Industry. Across the English Channel, France is poised on the brink of turmoil; the "divine right of kings" and the unquestioned authority of the church is about to yield to popular revolution and the Age of Reason. Beyond the broad waters of the Atlantic a new nation has been born, Europe's young and vigorous offspring, "conceived in liberty and dedicated to the proposition that all men are created equal." These world-shaking upheavals may have seemed remote to Playfair and Hall as they stood on the upright layers of ancient sandstone and let Hutton guide their imaginations into ever more remote epochs of the past. But what occurred at Siccar Point on that spring day in 1785 was no less consequential than what was happening in the new British mill towns, the streets of Paris, or Liberty Hall in Philadelphia. On the Berwickshire coast, human history and cosmic history were irrevocably wrenched apart. No longer would time tick off the course of human affairs; instead, human affairs became mere ticks in cosmic time.

But not immediately. The vistas of geologic time were too vast to be quickly grasped, the abyss too giddy, the implications for the meaning of human life too disconcerting. What had seemed plain to Hutton as he explored the hills and dales of Berwickshire, and to Playfair and Hall as they stood with Hutton at Siccar Point, needed to be hammered home by rhetorical and evidential persuasion. Playfair had his own role to play; his *Illustrations of the Huttonian Theory of the Earth,* published in 1802, was a translation of Hutton's labored prose into something more approachable by the common reader. But Hutton's central idea was too grand even for Playfair's lucid exposition. What was needed was an intellect of such down-to-earth power and subtlety as not to stand giddy in the face of events whose duration was measured in millions, or tens of millions, or hundreds of millions of years.

Charles Lyell (1797–1875), trained as a lawyer, was, like Mantell and Anning, irresistibly drawn to reading the book of the Earth, whose pages were increasingly laid open for perusal by the mines, quarries, railroad cuts, tunnels, and canals of the Industrial Revolution. Lyell took Hutton's idea of gradual change over eons of time and ran with it. His *Principles of Geology,* published in two volumes in 1830 and 1832, buttressed and amplified Hutton's uniformitarian philosophy, bringing to bear such a weight of evidence from his own observations in Britain and continental Europe that he is sometimes called "the Father of Geology." When it came to explaining the record of the rocks, Lyell got it pretty much right, and much of what he wrote remains valid today. The first volume of *Principles of Geology* came off the press just in time for young Charles Darwin to take it with him on his five-year, round-the-world voyage aboard HMS *Beagle.*

The decade preceding the publication of *Principles* was one of excitement among the fossilists of Britain. Gideon Mantell turned up at Cuckfield the femur (leg bone) of a land reptile that must have had a length (Buckland guessed) of sixty or seventy feet. The bones of other reptilian giants were disclosed at Stonesfield near Oxford. Mary Anning pried from the Lyme Regis shore the first complete skeleton of a long-necked plesiosaur (which today can be seen swimming across the wall of London's Natural History Museum). In 1830, the year of publication of volume 1 of Lyell's *Principles of Geology,* Anning found an ichthyosaur skull that measured five feet long! The learned gentlemen of the Geological Society in London enthused with each new find, and Buckland and Conybeare went to ever greater lengths to make the fossils proclaim the glory of God's handiwork during the supposed six days of creation. Lyell and his allies saw nothing in the record of the rocks except natural forces at work over eons of time—with no signs or evidences of divine interventions. The *uniformitarians* engaged in lively and sometimes bemused debate with their more religiously inspired colleagues. With each new fossil find, the world grew manifestly older. Soon, not even Buckland and Conybeare were prepared to defend the assertion by the seventeenth-century Bishop James Ussher, based upon the evidence of scriptures, that the world was created in 4004 B.C.

In October 1821, twenty-four-year-old Charles Lyell knocked on the door of Gideon Mantell's house on the high street in Lewes, mere feet from the prime meridian. He was at the threshold of his career as a geologist, and attracted by Mantell's confirmed reputation as a fossilist. The two men sat talking until the early morning hours, and a lasting friendship was established; they became firm allies in the battle to wrest time from the theologians. A print used as the frontispiece of Mantell's *Illustrations of the Geology of Sussex* (1827) depicts a visit made by Mantell, Lyell, and Buckland to the quarry at Cuckfield which had yielded up Mantell's iguanodon tooth and many other fossils. (See figure 3-4.) It is a rainy afternoon in March of 1825. The three geologists are in top hats and gentlemanly garb. They are accompanied by half a dozen less formally attired quarry workmen. Mantell is presumably the person at right, standing behind a vertical slab of sandstone etched with a fossil fern. Lyell or Buckland wields a hammer to release a reptilian bone from the rock. In the background is the spire of Cuckfield Church. (The quarry has since been filled in, and a cricket field stands in its place.) It is a picture that lends itself to metaphorical interpretation. Whatever the differences between Lyell and Mantell, on the one hand, and Buckland, on the other, it is clear that the story of the past will ultimately depend upon the evidence of the fossiliferous strata and not upon the authority and

FIGURE 3-4. Gideon Mantell, Charles Lyell, and William Buckland in the Cuckfield quarry, 1825.

tradition represented by the distant church spire. I particularly like the conjunction at the right of fossil ferns and living plants, reflecting each other mirrorlike across eons of geologic time. *The present is the key to the past,* said Hutton and Lyell; if we want to understand the history of the Earth and its denizens, let us look to natural forces at work on the globe today and apply them to the past.

Mantell's journal entry for May 21, 1831, recounts another expedition he made with Lyell to a quarry at Horsham, at which time they happily examined slabs of ancient sandstone covered with ripple marks. Mantell subsequently described these sandstone surfaces in a note to the *Edinburgh New Philosophical Journal.*

No one who has observed the action of waves on a living beach can doubt that the *identical undulations* in ancient sandstones were made by the same agency, asserted Mantell. Then he penned a line that could stand as the epigraph of this book or any book on the history of science: "Obvious as the cause of this curious appearance seems to be, yet it has been a subject of dispute among men of science, the mind being but too apt to seek for a mysterious agent, to explain effects which have been, and are still being, produced by some simple operation of nature."

What that "mysterious agent" might be has varied from time to time and place to place, but invariably it has taken the form of a humanlike creator, or animal creator with human qualities. Jean Piaget has shown us that young children invariably evoke artificialist explanations of natural phenomena, so that even Sun, Moon, wind, and clouds become the products of a conscious agency who has acted with particular reference to the child. Likewise, anthropologists see artificialist explanations at work within so-called primitive cultures around the globe. Middle Eastern creation myths, such as that recounted in Genesis, are among common examples of artificialist tradition. The tendency to understand natural phenomena in a way that emphasizes the centrality of human life is very strong indeed. Not even scientists are immune to the seductive attractions of animism and artificialism.

Scholars of mythical thinking and early religion, such as Mircea Eliade and Joseph Campbell, have shown that the concept of time in traditional cultures is based upon recurring cycles rather than linear succession. The recurring cycles are closely associated with celestial phenomena—the lunar and especially solar cycles that impact so forcefully upon every aspect of traditional human life. *Traditional time is the eternal repetition of an unvarying cosmic rhythm.* Repetition and simultaneity are of more significance in traditional cultures than duration and directionality. In traditional cultures, "an object or an act is 'real' only in so far as it imitates or repeats an ideal prototype," writes the mathematical philosopher G. J. Whitrow; in other words, the past is key to the present. "We are, therefore, presented with the paradoxical situation that in his first conscious awareness of time man instinctively sought to transcend, or abolish, time," says Whitrow. Our ancestors, it seems, sought to live in an eternal present. A present, it hardly needs to be said, that is centered on self.

Religious and sacrificial rites in traditional cultures are performed at specific moments in the solar or lunar cycles; prayers are recited at designated hours of the day. Every ceremonial act is a repetition and celebration of an archetypal divine act. This is the "eternal return" described so persuasively by Eliade and Campbell. "Even in medieval Europe the first stages in the development of the mechanical clock seem to have been influenced by the monastic demand for accurate determination of the hours when the various prayers should be said, rather than for any desire to register the passage of time," writes Whitrow. Across the globe, among cultures as diverse as those of the Australian aboriginal peoples and the Egyptians of pharaonic times, only the *changeless* is significant. According to all traditionalist cultures, the world came forth complete and perfect from the hands of an artificialist creator, and if we wish to understand contemporary events or phenomena we must make reference to the creator's intent at the time of creation.

And so it was that the culture in which Hutton and Lyell found themselves was that of Judeo-Christian tradition, in which the myth of the eternal return had been elevated to a single great cycle that encompasses the creation and fall, the coming of the Savior, his sacrificial murder, and—in the fullness of time—his triumphal return. This is the great arc of sacred history into which Conybeare and Buckland sought to force the testimony of the rocks. Ripple marks on ancient sandstone slabs bore witness to a particular chapter of this human-centered story, they asserted—God's singular destruction of the world by flood as recorded in scriptures—and not to the slip-slap of wave on sand such as occurs on every beach of the world today. How radical, then, of Playfair to examine the contorted strata of Siccar Point and proclaim, "The result . . . of this physical inquiry is that we find no vestige of a beginning, no prospect of an end." History, in James Hutton's view, is not an eternal present made significant by a singular and complete act of past creation, but an apparently endless succession of essentially indistinguishable moments—one damn thing after another—not a thrilling prospect for people who considered themselves the central point and purpose of it all. No wonder even scientists of Hutton's time were conflicted about the meaning of fossil bones and ripple marks in sandstone.

Gideon Mantell's hometown of Lewes sits in a notch in the chalk downs where the River Ouse cuts its way to the sea. High white escarpments rise up to the east, west, and south; it was by cataloging the fossils of the chalk that Dr. Mantell established his reputation as a geologist. Approach England from France across the narrowest part of the English Channel, and you cannot help but be dazzled—especially on a sunny day—by the brightness of the vertical cliffs of soft white rock that rise up like battlements.

This is the prospect that gave England its poetic name, Albion, from *albus*, the Latin word for "white."

We know much more today than did Mantell about the origin of the chalk stratum, which rises to the surface over much of southeastern England (and also across the Channel in France). The chalk is almost completely pure calcium carbonate, in some places more than one thousand feet thick. Microscopic examination reveals that it is composed of tiny skeletal plates of marine algae called coccospheres, which apparently bloomed as an oozy mass on the floor of a rather shallow saltwater sea about 100 million years ago. Modern calculation suggests that the rate of accumulation of sediment was about one inch every one thousand years, so one can easily grasp the eons required to deposit beds of chalk one thousand feet thick. Within these strata Mantell found the fossil remains of fish, ammonites, sea urchins, and dozens of other marine creatures, conclusive evidence of a time when a warm-water sea lay upon much of what is now northeastern Europe.

But Mantell had no notion of rates of sedimentation or of the numbers of years recorded in his sequence of fossils, only that a very long time indeed was evidenced in the rocks. Nor did his understanding of the strata yet extend much beyond his native Sussex. It would take someone with a more *expansive* view of England's geology to unravel the riddle of the chalk.

I know in advance what sorts of rocks I will encounter as I make my way northward along the prime meridian because I have long owned the beautiful two-sheet, wall-sized geologic map of Britain published by the British Geological Survey, and many times I have made this very walk *in my imagination.* As I descend from the South Downs near Lewes, the chalk I have been walking across since Peacehaven gives way to sandstone, the so-called Upper Greensand. As I progress northward across the broad vale called the Weald, Upper Greensand yields to Lower Greensand, then clay, then a third variety of sandstone again at the hilly central Weald. Now something interesting happens as I continue northward toward London: the sequence of strata is reversed. Clay, Lower Greensand, Upper Greensand, and then at the south-facing escarpment of the North Downs chalk again makes its appearance. From the high points of this escarpment, looking north on a clear day, one can see the towers of London in the distant valley of the Thames. As I proceed along the gentle descent to London, through the southern suburbs of the city, the chalk gives way to a sequence of clays and mudstones. I cross the river at Greenwich (a pedestrian tunnel facilitates my passage), and the clays and mudstones *reverse* their sequence as in a mirror image. As I move away from the northern suburbs I encounter chalk again, and then, near Cambridge, the same sequence of Upper Greensand, Lower Greensand, clay, and sandstone that I saw in the Weald.

How to account for this rhythm in the rocks? I have posed this puzzle for my students, and after a bit of head-scratching they figure out a solution. All of southeastern England is underlain by the same sequence of sedimentary rocks, with the oldest strata exposed at the central Weald and the youngest at London. The layers of rock have been gently folded and eroded so that today different strata are exposed at the surface. (See figure 3-5.) Of course, having the modern geologic map of Britain hanging on the wall of the classroom helps solve the puzzle. A solution was not so obvious when there was no countrywide geologic map available. The man who first conceived such a map was the self-educated son of a blacksmith, William Smith (1769–1839).

Smith was born in the village of Churchill in Oxfordshire in the year that James Watt was granted a patent for the condensing

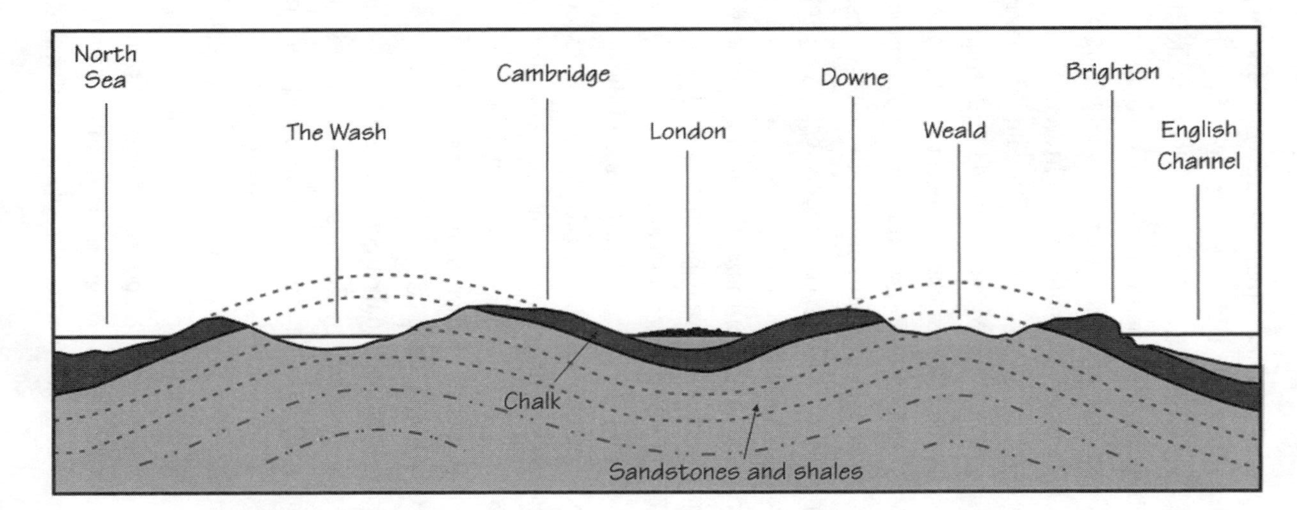

FIGURE 3-5. An idealized representation of the subsurface geology of southeastern England.

steam engine, an event that might be taken as the beginning of the Industrial Revolution. He died not long after Charles Lyell published *Principles of Geology.* His life aptly illustrates the intimate connection between the industrialization of Britain and the birth of geology as a science. During the seventy years of Smith's lifetime Britain changed dramatically and irreversibly. Population soared, but also shifted, away from tiny agricultural villages sprinkled across the countryside to big, crowded industrial centers in the Midlands, southern Wales, the northeast coast near Newcastle, and the region around Glasgow in Scotland. A comparison of a population density map of Britain for the mid-1800s with a geologic map shows a striking coincidence between people and coal. Within a few generations, a nation of farmers became a nation of coal extractors and burners. Powered machinery did the work of the hand: spinning, weaving, forging, and pumping water from mines. The hiss and rattle of steam locomotives replaced the call of the cuckoo as the characteristic sound of the countryside. Only the great commercial center of London is an exception to the coincidence of population and coal, but canals and then railroads were built to bring coal to the capital for the heating of homes and businesses, together with the useful products of the industrial centers in the Midlands.

Smith began his professional career as a surveyor and engineer in the coalfields of northern Somerset between the towns of Wells and Bath. The coal here was plentiful, and Somerset was not far removed from cities such as Bristol, Southampton, and London, but the coal was not exposed at the surface as in more favored places such South Wales and the Midlands. The coal was therefore expensive to extract, and if the Somerset mines were to remain viable in the face of competition from farther afield they must get their product to market cheaply. A canal was required, linking the Somerset collieries with the other canals that bound the country together. Previously, in Oxfordshire, Smith had worked as an apprentice surveyor and had mastered the skills of that trade. He now found himself in the right place at the right time to be given responsibility for planning the course of the Somerset canal.

The insatiably curious Smith, who had collected fossils and rock specimens as a child, did not waste this opportunity to observe the *inside* of the Earth. He descended mines and recorded in the walls of vertical shafts successive layers of rock—sandstones, mudstones, coal—all well known to the miners but not yet described by science. Along the route of his new canal he noted where these underground strata intersected the surface and realized that he could track gently rising and falling layers of rock across entire counties, indeed across the nation, *in three dimensions.* And he had one great insight that was destined to transform geology and the story of life on Earth: *Each layer of sedimentary rock has its own unique assembly of fossils.* As one ascends through the strata, from the lowest and therefore presumably oldest rocks to the uppermost and therefore younger rocks, the fossil organisms the rocks contain become more complex and more like forms of life that exist on Earth today.

Smith was by no means the only person in Britain interested in fossils. From the mid-1700s until the mid-1800s fossil collecting was something of a national craze, and many households *of a certain class* possessed impressive troves. Smith's friend Reverend Benjamin Richardson was a typical amateur fossilist. At his home in Bath, he had amassed a glorious assembly, which, typically, he had arranged by type of fossil: all the ammonites on one shelf, all the crinoids on another, and so on. Smith took one look at his friend's collection and volunteered to arrange it in proper stratigraphic sequence, with the oldest fossil assemblies on the bottom and the youngest on the top. He did so, to Richardson's amazement.

Smith had no idea why fossils varied from strata to strata. Although a theological agnostic, he had no idea of evolution. What was important to him was not *why* the fossils differed, but how the differences might be used to identify strata across wide geographic areas. As the eighteenth century gave way to the nineteenth, he conceived of the project that would become the crowning achievement of his life: a geologic map of Britain, on which every kind of rock exposed at the surface would be identified by key and color.

The first version of Smith's completed map, roughly six by nine feet, printed on fifteen separate sheets, appeared in 1815, titled *A Delineation of the Strata of England and Wales with part of Scotland; exhibiting the Collieries and Mines; the Marshes and Fen Lands originally Overflowed by the Sea; and the Varieties of Soil according to the Variations in the Sub Strata; illustrated by the Most Descriptive Names. By W. Smith. Aug*[st] *1, 1815.* (See figure 3-6.) Four hundred copies of the map were printed, and about forty are known to remain, each immensely valuable. I could not hope to possess an original, but a poster reproduction of Smith's map is available from the British Geological Survey, and this I do own. It is instructive to place Smith's map next to the modern geologic map of Britain. The coincidence is striking. In the nearly two centuries that have passed since Smith published his map, thousands of geologists have tramped every square inch of Britain, recording fine details of geology. Yet when the maps are displayed side by side, one realizes that Smith, all by himself, got it mostly right the first time out.

With Smith's map or the modern map it is easy to work out the three-dimensional disposition of strata along my meridian walk,

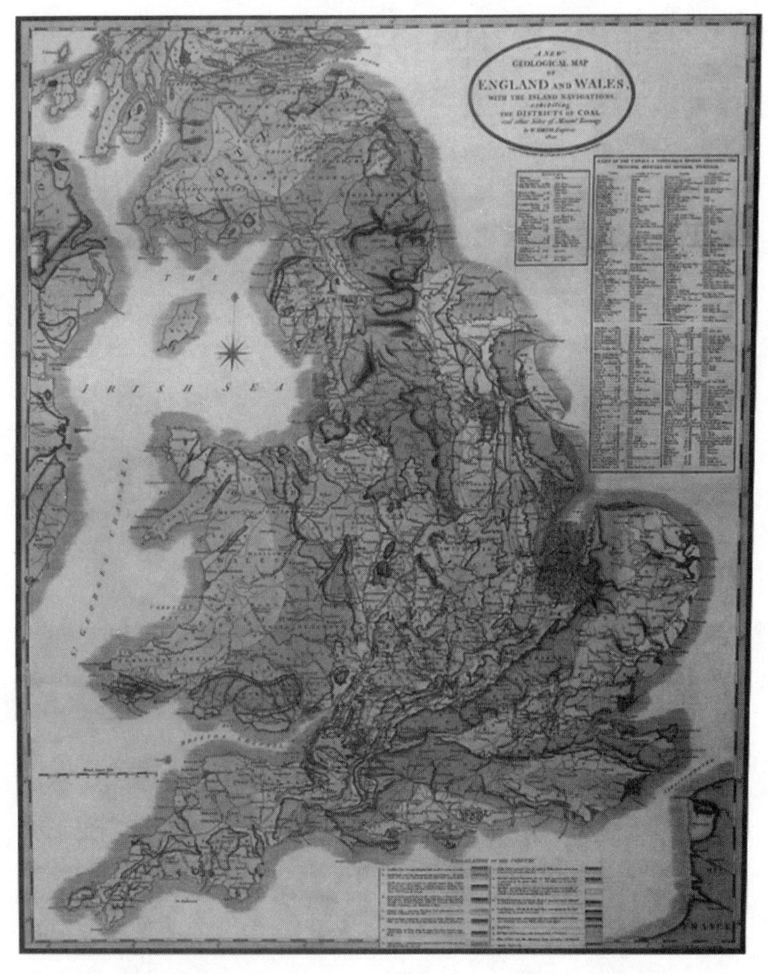

FIGURE 3-6. William Smith's *A New Geological Map of England and Wales,* 1820.

and, following the principles of Hutton and Lyell, to deduce something of the general history of this part of the Earth's surface. The chalks, sandstones, muds, and clays were laid down in horizontal beds at times in the past when this part of the world was submerged by sea or nearly so. The oldest strata, as evidenced by the least complex assemblies of fossils, are the sandstones and mudstones of the central Weald and the region north of Cambridge. As these beds were deposited in subsiding basins during Jurassic and early Cretaceous times (200 to 140 million years ago), the London area stood above the sea and supplied terrestrial sediments to the basins. Later, differential subsidence ceased, and the entire region became covered with a warm shallow sea in which the chalk was deposited. Subsequently, the region was gently squeezed and folded along an east-west axis. Those parts of the ancient strata that were folded upward (the Weald chalks, for example) were eroded away. Those parts thrust downward (the London area or the English Channel) received even younger sediments. And so today, while walking from Peacehaven to Cambridge along the prime meridian, I am making a journey backward, then forward, then backward again in geologic time, a trek of some days in space but of hundreds of million years in time.

William Smith had no idea of the number of years required for the deposition of the rocks he so accurately mapped, nor did he comprehend the details of England's paleogeography, but he certainly understood the general story of the past and saw as deeply into the abyss of time as did Playfair and Hall when they stood with Hutton at Siccar Point. For his persistence in gathering geologic data and compiling and publishing his map, Smith at first reaped nothing but troubles. As the son of a blacksmith in class-ridden England, he was generally snubbed by the gentlemanly elite of London's Geological Society. The success of his great map was undercut by a competing map published by the Geological Society, largely plagiarized from Smith. His financial affairs went from bad to worse, and for a time he ended up in debtor's prison.

But all turned out right in the end. Along with the Industrial

Revolution (and revolutionary ideas from across the Channel) came the ascendancy of a new, practical-minded middle class of men and women who did not look down their noses on the likes of Smith. In his old age, the man who had come to be nicknamed "Strata" Smith was called out of obscurity to London and awarded the Geological Society's first Wollaston Medal, the highest honor the society bestows. (Today the medal is the equivalent of a Nobel Prize in geology.) The year of Smith's award was 1831, and in another part of England a young son of the new entrepreneurial class, Charles Darwin, having shown no aptitude for a career as a physician or man of the cloth, was wondering earnestly what to do with his life.

4

ANTIQUITY OF MAN

When Charles and Emma Darwin bought the house that would be their family home for forty years, at Downe, sixteen miles south of London, one of Charles's earliest improvements was to have the flints removed from the property's chalky meadow. Tramp across any plowed field in England's chalky North or South Downs, and these fist-sized nodules of pure, hard, yellow silica (silicon dioxide, better known as quartz) are common underfoot. In the cliffs beside the English Channel at Brighton I saw them interspersed as dark layers in the chalk. In the Downs, I stumbled over them along the path. The flints are chemically very different from the chalk, and their presence in the otherwise pure calcium carbonate has long been something of a geologic mystery. The most plausible modern explanation is that the silica nodules had their origin in siliceous sponges that grew on the sea floor and other siliceous marine microfossils. When these organisms died, their substance dissolved in sea water and was dispersed within the carbonate ooze, then precipitated out around other organic remains in a process called petrification. Petrified sea urchins, for example, can be found among the flint nodules of plowed fields. For Darwin, these hard stones were more than an agricultural nuisance; they were a puzzle to be solved. Today, his home has been lovingly restored, and a collection of flints is arrayed on a table in

Darwin's study, as they might have been when Darwin sat beside them pondering their meaning.

The house is called Down House, after the village (which added an *e* to its name shortly before the Darwins moved in). Few homes are more intimately bound up with the life work of a resident, and my visit there (at longitude 0° 3.4") on my meridian walk was a pilgrimage to one of the shrines of science. Here Darwin lived and worked, in the midst of a large and happy family, in relative seclusion from the hustle and bustle of scientific London. All his life—subsequent to his return from his five-year circumnavigation of the Earth aboard HMS *Beagle*—he suffered from a mysterious debilitating malady which may have been partly psychosomatic; nevertheless, he accomplished more than do most people who boast good health. His home, garden, greenhouse, dovecote, and the land around were his laboratory, and here he assiduously gathered evidence to buttress the one great idea that had taken root in his mind during the *Beagle* voyage: the transmutation of organisms by natural selection.

The idea of natural selection was not original with Darwin. His grandfather Erasmus Darwin, among others, had toyed with the idea of transmutating organisms. Then, in 1858, after working on his theory for twenty years, Charles received a copy of an essay by the naturalist Alfred Russel Wallace advancing a hypothesis identical to Darwin's own. This shattering and unexpected development prompted Darwin to hurriedly bring his great work into print. *On the Origin of Species by means of Natural Selection or the Preservation of Favoured Races in the Struggle for Life* was published in November 1859, to instant scientific acclaim and public notoriety. Although the central idea was identical to that of Wallace, so forcefully did Darwin amass evidence for evolution by natural selection that the idea has quite properly come to be known as "Darwinism."

Darwin achieved for biology what Hutton, Smith, and Lyell had done for geology: He provided a natural mechanism to account for phenomena that had hitherto been explained by the artificialist interventions of a divine being. The theory is simplicity itself: (1) Species are variable; (2) variations are maintained during reproduction; (3) individuals produce more offspring than are needed for the species to survive; (4) those individuals that are well adapted to their environment will be more likely to survive and reproduce, thereby passing on their traits to succeeding generations. In fact, so elegantly simple is the idea of evolution by natural selection that Darwin's friend and champion Thomas Huxley, among others presumably, wondered why he had not thought of it himself.

Why Darwin? Even as a young man Charles seems to have been gifted with a capacity to observe without preconceptions. He flirted briefly with becoming a doctor, like his father and older brother, but found he had no stomach for observing pain. (In the days before anesthetics doctors needed to be hardened to the agonies of their patients.) His innate sympathy with living beings served him well as an observer on the *Beagle* voyage. Young Darwin had also considered becoming a man of the cloth, but his growing doubts about Christianity dissuaded him from that course. His detachment from Christian dogma reinforced his skill as an unbiased observer of the natural world. When the opportunity came to be resident naturalist on the five-year voyage of the *Beagle,* Darwin jumped at the chance. It would be the defining experience of his life. Everything he saw on the voyage, he saw afresh, with a mind that was awake and aware of new possibilities.

Of course, a necessary prerequisite for natural selection to work is time—lots and lots of time—and this Darwin was supplied by the geologists. Charles Lyell, author of the book Darwin took with him aboard the *Beagle*, later became his friend and confidant. Down House was wonderfully situated to serve Darwin's interest in geology. He was certainly aware of how William Smith and others accounted for the folded strata of southeastern England, and he knew of Gideon Mantell's fossil finds. There is a moment in Darwin's great book, *The Origin of Species*, when he describes one of his favorite walks, to the edge of the chalk escarpment a few miles south of Down House, with a view far out over the Weald toward the mirroring escarpment of the South Downs twenty miles away. Standing there, he could imagine the huge uplifted arch of chalk and sandstone that must have been eroded away to create this wide vale (see figure 3-5), all during relatively recent geologic times. The vanished strata might have been 1,100 feet thick, he reckoned, and at present rates of erosion it would have taken 300 million years for water and weather to eat away the rocks. He wrote: "I have made these few remarks because it is highly important for us to gain some notion, however imperfect, of the lapse of years. During each of these years, over the whole world, the land and the water have been peopled by hosts of living forms. What an infinite number of generations, which the mind cannot grasp, must have succeeded each other in the long roll of years."

After my own visit to Down House I walked the country lanes and footpaths to the place where Darwin might have stood at the lip of the North Downs, and let my own imagination reconstruct the great dome of rock that once stood in place of the valley that lay before me. Thanks to Darwin and his contemporaries, my mind could readily grasp "the long roll of years." And thanks to a recent feat of civil engineering—the Channel Tunnel, or Chunnel—I could imagine more vividly than could Darwin the way the chalk dips below the English Channel to reemerge in France. The lowest layer of these strata is a seventy-five-foot-thick clayey chalk marl, a perfect material for tunneling: strong, impermeable to water, and easily penetrated by boring machines. The Chunnel makes its way from England to France almost entirely within this strata, dipping two hundred feet below the sea floor to follow the gently folded rock, confirming *by being there* what Smith, Lyell, and Darwin saw only in their mind's eye.

When Darwin returned from his voyage on the *Beagle,* at age twenty-seven, he considered in his methodical way the pros and cons of married life. He settled finally on the pros and proposed to Emma Wedgwood, his first cousin and childhood companion. During their engagement, Darwin told his pious fiancée of his growing doubts regarding Christian revelation. He had already seen enough evidences of ancient life to know that the world was older than the thousands of years allotted by Genesis, and he had seen enough inherent cruelty in nature to doubt the existence of an all-powerful loving God. He doubted, too, the promise of an afterlife.

Emma's religion was an affair of the heart, not the intellect. The hardest thing for her to bear was the possibility that Charles, by his doubts, had forfeited their chance of being reunited in heaven. Throughout their married life, their religious differences lay like a dark shadow between them, but each respected the other's beliefs. Together they had ten children. (The third, a daughter, lived only three weeks.) In 1851 Darwin's eldest daughter and treasured child, Annie, died at the age of ten of what is presumed to have been tuberculosis. During Annie's illness, Darwin was at her bedside night and day. Her death gave poignant meaning to his developing notions of the amorality of nature and the struggle of all creatures for survival.

Annie's death was a test of Emma's faith and of Charles's

doubts. A widely held view among Christians at that time was that death is due to sin—either the victim's, or another person's, or Adam's. Most assuredly Emma did not blame Annie. If she thought Charles's apostasy was implicated, she did not say so. Since God cannot cause evil, she assumed that Annie's death must be meant for good in some mysterious way. Charles did not believe there was any divine purpose behind Annie's death. For him, death was a purely natural process, part of the machinery of life that drove evolution toward "endless forms most beautiful." The only comfort he had at Annie's death was that during her brief life he had never spoken a harsh word to her. He was distressed that *he* might be responsible for her death, not because of his theological doubts but by heredity; he was sickly all his life.

Humans are animals, Darwin believed, and like all animals they are locked in a struggle for existence, which, left to itself, eliminates the weak. Twenty-six years after Annie's death, Dr. Robert Koch took the first photograph ever published of a bacterium, the tuberculosis pathogen, and so confirmed the germ theory of disease. As Charles had guessed, Annie had died so that another creature might live. But humans can escape from the relentless logic of natural selection, Darwin firmly believed. By caring lovingly for the sick and weak, we lift ourselves *above* our animal natures.

Darwin's attendance at Annie's bedside was unwavering. He never doubted our responsibility to cherish the least advantaged—"the noblest part of our nature," he called it—and he strongly opposed what came to be called "social Darwinism," the natural rule of the strong in human affairs. After his daughter's death, he put the notion of a loving God firmly and fixedly behind him. The Creator he found in nature was henceforth "a shadowy, inscrutable and ruthless figure." Darwin himself was far from shadowy, inscrutable, and ruthless. He was open, forthright, and kindly, and even in his grievous bereavement he continued to see "the face of nature bright with gladness."

Walk north from Darwin's village of Downe, and the land falls gently away toward London and the valley of the River Thames. The chalk strata dip below London more steeply than the surface topography, and the natural bowl formed by the downward folded chalk is filled with the more recent muds and clays of the London Basin. The chalk rises again to the surface north and east of the city where it finds its greatest surface exposure in the counties of Suffolk and Norfolk.

In the summer of 1868, nearly a decade after the publication of *The Origin of Species*, the British Association for the Advancement of Science held its annual meeting in the town of Norwich, ninety miles northeast of London, just at the edge of the Norfolk chalk. At that meeting Thomas Henry Huxley (1825–95), one of the greatest natural philosophers of his day and a scrappy champion of Darwin's new theory of evolution by natural selection, delivered a talk titled "On a Piece of Chalk." His audience were the ordinary workingmen of the town, and his subject was engagingly simple. The town is built upon the same beds of soft, white rock that rise and fall along my meridian walk with stately rhythm. Some of the carpenters in the audience may have carried a lump of Norwich chalk in their pockets.

From a piece of chalk that he held in his hand Huxley extracted an astonishing story of a vast saltwater sea that once lay upon Britain and of the microscopic creatures that lived in that sea in prodigious numbers. These tiny animals contributed their calcareous skeletons to bottom sediments that were ultimately compacted into chalk, Huxley told his audience. The microscopic skeletons, of a wonderful geometric complexity, are often beautifully preserved.

Eleven years earlier, the British Admiralty had commissioned Huxley's friend Captain Joseph Dayman to sound the floor of the Atlantic Ocean along the route of the proposed Atlantic Cable. Dayman sailed from Valentia, Ireland, to Trinity Bay in Newfoundland, measuring the depth of the sea and retrieving samples of mud from the ocean bottom. These specimens of deep-sea sediments were submitted to Huxley for scientific examination. Huxley assured his Norwich audience that the sediments brought up from the floor of the present ocean contain the same sorts of microscopic organisms that are preserved in the Norwich chalk. Where you see the same effect, he said, it is reasonable to assume the same cause. If the fossils in the Norwich chalk resemble the creatures found in the muddy depths of present ocean—and nowhere else in the world—it is reasonable to assume that the chalk was once sea bottom sediments.

The chalk beds at Norwich are hundreds of feet thick. "I think you will agree," Huxley told his audience, "that it must have taken some time for the skeletons of animalcules of a hundredth of an inch in diameter to heap up such a mass as that." How long? Embedded within the chalk are the fossils of higher animals, corals, brachiopods, sea urchins, and starfishes, altogether more than three thousand distinct species of aquatic animals. Among these fossils are some curious combinations: for example, a coral-covered shellfish affixed to a sea urchin. Here was a hint to the age of the chalk sea, and Huxley unraveled the story: *The sea urchin lived from youth to maturity on the floor of the sea, then died and lost its spines, which were carried away. The shellfish adhered to the bared shell and grew and perished in its turn. Finally, coral-building organisms covered both shellfish and urchin, lived out their lives, and expired. And all of this unfolded before slowly accumulating sediments encased these creatures in an inch or two of chalky mud.* It was easy for Huxley to deduce that a minimum of tens of thousands of years was required for the deposition of chalk beds hundreds of feet thick.

But Huxley's story of the great abyss of time was not yet complete. Where the River Yare flows through Norwich it cuts down through sandy clays to expose the chalk. Since the clays lie above the chalk, they must have been deposited at a later time. Between the clay and the chalk there is a layer of vegetative matter, including the fossilized stumps of trees standing as they grew—fir trees with their cones and hazel bushes with their nuts. Clearly the chalk must have been uplifted from the floor of the sea before forests could grow upon it. A greater surprise! Among the bolls of the trees are the fossilized bones of elephants, rhinoceroses, hippopotamuses, and other wild beasts that roamed the ancient forest. And above the forest beds, interspersed within clay of a marine origin, are the fossils of walruses and other cold-water sea creatures now found only in the icy waters of the North.

Sea, land, sea, and now land again! Spectacular changes in climate. What forces caused such giddy transformations? Huxley did not know and readily confessed his ignorance, and in this—his willingness to say "I don't know"—he embodied the modern scientific spirit. But he *did* know that the evidence of the Norwich rocks "compels you to believe that the earth, from the time of the chalk to the present day, has been the theater of a series of changes as vast in their amount as they were slow in their progress."

How astonished must have been the members of the Workingmen's Association of Norwich to hear of this dramatic extension of the history of their town. Huxley's lecture "On a Piece of Chalk" has come down to us as a little classic of scientific exposition, as engaging and informative today as in 1868. He did not talk down to his audience. He did not grind any theological axes. He simply directed the attention of his listeners to the rocks and let the rocks speak for themselves.

Huxley and Darwin admired each other greatly, but it would be hard to find two men more unlike. Darwin was reclusive; he abhorred controversy. Anticipation of the furor that his great book would cause may have been one reason he delayed publication; it may even have aggravated his sickliness. Huxley, by contrast, loved a scrap. It was Huxley who took on Bishop Samuel ("Soapy Sam") Wilberforce at the 1860 Oxford meeting of the British Association for the Advancement of Science. Darwin's just published book was the talk of that gathering. Although Darwin had left humans out of *The Origin of Species,* he did suggest briefly in the book's conclusion that "much light will be thrown on the origin of Man and his history." The implication of this teasing throwaway line was clear enough, and Wilberforce led the charge of the established order against the upstart idea that humans might be less than a unique and special creation of God. At one point in his remarks, the bishop of Oxford turned to Huxley, who represented Darwin on the dais, and asked "whether it was on his grandfather's or his grandmother's side that he was descended from an ape." It was just the opening Huxley needed, and he rose to suggest that he would rather have as his ancestor a miserable ape than a man who would use his great faculties and influence for the purpose of ridicule. The exchange brought the house down and won for Huxley the nickname "Darwin's Bulldog."

While Huxley carried the battle forth, Darwin remained secluded at Down House, publishing works on the fertilization of orchids by insects, the movements of climbing plants, and the variation of plants and animals under domestication. At last, he embarked upon the long postponed but necessary project, the application of the principles of natural selection to human evolution. *The Descent of Man* was published in 1871 to renewed public controversy, although by then most scientists had been won over to Darwin's view.

More appeared to be at stake than mere science. The *Times of London* put it succinctly: "If our humanity be merely the natural product of the modified faculties of the brutes, most earnest-minded men will be compelled to give up those motives by which they have attempted to live noble and virtuous lives, as founded on a mistake." The *Times* was correct that men and women of enlightened scientific learning might question their previous motives for goodness—anticipation of the rewards of heaven or fear of hell—but the paper was wrong in believing that only churchgoing Christians might choose to lead noble and virtuous lives. It was churchgoing Christians, after all, who initiated and sustained the African slave trade, the greatest moral abomination of the century. Darwin and Huxley were both agnostic, yet both men found ample motives to live virtuous lives, and both were firmly on the side of progressive social and political change.

Of course, it was progressive social and political change that the *Times of London* feared. The mid-nineteenth century was a time of social upheaval throughout the European continent. Old class structures were being dismantled, inherited privilege wrested away, the power of monarchies and churches challenged. In England the established powers felt under siege. The new working class, crowded into factory towns, seethed with unrest. The Anglican Church, that great prop of crown and privilege, understood well enough that once the divine origin of the existing order was questioned, the whole house of cards would come tumbling down.

The philosophical buttress for the ancient social order was the so-called Great Chain of Being, a supposed ladder of creature-dom that reached from the foot of God's throne down through the angelic realm to humans—who stood at the apex of material existence—and farther down through the various species of animals, plants, and inanimate substances to the the dregs of the Earth. Every creature had a preordained place on the ladder. Kings ruled by divine right. Privilege came from the hand of God. The sons of blacksmiths were *meant* to be sons of blacksmiths, as William Smith discovered to his misfortune, and woe to the son of a blacksmith who sought to hobnob with the London elite. Mary Anning was preordained *by her sex* to surrender scientific precedence and privilege to men. Of course, the entire system was riddled with hypocrisy; William Smith's great foe in the Geological Society was George Bellas Greenough, a wealthy dandy who became the first president of the society, and who took pains to conceal the fact that his grandfather had made the family fortune selling a quack remedy known as Greenough's Liver Pills.

And now along came Darwin to call the ladder of established privilege into question. Creation is not a vertical chain of cast-iron links, he said, but a living, evolving family tree. The place of any organism in nature is not ordained by divine fiat; the web of creaturedom is woven and rewoven by the action of natural agencies. The establishment, of course, would have none of it. Darwin's onetime professor of geology at Cambridge, the Reverend Adam Sedgwick, went so far as to blame the socially leveling French Revolution for such "gross (and I dare say, filthy) views" as transmutation of species. Cambridge and Oxford were bastions of Anglican privilege; allow species to change, wrote Sedgwick, and you "undermine the whole moral and social fabric [bringing] discord and deadly mischief in its train." Even Darwin knew his ideas were politically subversive. At some deep emotional level he might have felt a traitor to his own relatively privileged class, and this, as much as anything, may have been the source of his physical torment.

When Huxley lectured to the workingmen of Norwich, he undoubtedly considered himself as something of an evangelist bringing a new gospel to the masses. But he was not afraid that an informed rabble would rise up and loose chaos upon the land. With the transmission of scientific learning to the masses he looked for an improvement in morals and a universal reign of liberty, equality, and fraternity, a new heaven and a new Earth. Of course, in the long run, like all of us, Huxley struggled to make sense of a world that could sometimes be viciously cruel. As with Darwin, his faith in the consolations of scientific inquiry was sorely tested by the death of a beloved daughter. Family tragedy, and the social and economic upheavals of the day, sometimes led him to the brink of despair, with no promise of a happy afterlife to steady his course. The advancement of science, so dear to his heart, provided scant solace in times of personal trial, nor did it usher in the era of universal peace and prosperity that he so earnestly hoped for. Thomas Henry Huxley was learning a lesson that must be relearned by every generation: To admit ignorance, to face the mystery of existence on one's own two feet without the prop of true belief, requires courage, more courage than many of us are willing or able to muster.

"You darkness, that I come from, I love you more than all the fires that fence in the world," wrote the poet Rainer Maria Rilke. It is a curious thought: that one should love darkness more than light.

Most of us want the story of our origins to be crystal clear, illuminated by a certain light, which is why, I suppose, so many of us cling with fierce determination to the artificialist creation myths of our ancestors. Darwin, Huxley, and their confrères challenged the ancient stories and offered in their place—what? Essentially, darkness. Their case for human evolution by natural selection rested mostly upon circumstantial evidence. No divine revelation confirmed its veracity. The implication of Darwin's *Descent of Man* was that humans share an ancestor with the apes—Bishop Wilberforce's bête noire—but who or what that ancestor might be was lost in the abyss of time.

In spite of our hankerings for specialness, the progress of biology since Darwin's time has made one thing irresistibly clear: We are mired in nature up to our necks. A zillion bacteria inhabit our guts. Mites creep in the forests of our eyelashes. Viruses swim in our blood. We depend utterly upon plants for energy from the sun. An uncountable number of microscopic marine organisms maintain the air we breathe. All of this is clear. But what of the darkness, the 4 billion years of hidden history—the patient crafting of complexity, the long unfolding of diversity? According to geneticists, every cell of our bodies remembers the eons: We share most of our DNA with other primates, and a lot of it with bugs and barnacles.

We are related to every organism on Earth by common descent, said Darwin. So we look for the evidence. We split open sedimentary rocks along their seams and spill fossils into the light after millions, or billions, of years of darkness. Like hieroglyphics on the walls of a newly opened Egyptian tomb, the fossils are a record of our past, and the record is more detailed than many people realize. Certain dragonfly fossils from 200-million-year-old Jurassic limestone show every vein of the finely netted wings. A juvenile theropod dinosaur from Cretaceous limestone deposits in southern Italy has exceptional preservation of soft tissue: muscles, gut, possibly traces of liver. Chinese and American paleontologists have found what are purported to be microscopic animal embryos from 570-million-year-old phosphate deposits of southern China, fertilized eggs in the earliest stages of cell division: one cell, then two, four, eight, sixteen, and so on.

But the record of the fossils is inevitably incomplete. The crust of the Earth is like a great encyclopedia of which only a few pages have been opened. "The past is continually erased, and the record of the most distant time survives only by a chain of minor miracles," writes the paleontologist Richard Fortey. The Italian theropod dinosaur, for example, apparently expired in a shallow lagoon and was quickly covered with a fine-grained mud low in oxygen, two requirements for detailed preservation of soft tissue.

Few chapters of the Earth's past are as fully recorded as are the Jurassic seas explored by Mary Anning. (In London's Natural History Museum, where many of her fossils reside, one almost feels as if one is swimming in those ancient seas.) Darwin insisted that humans are related to Anning's ichthyosaurs and plesiosaurs, although distantly. More disturbingly, he supposed other primates—chimps and orangutans—are our close cousins. Somewhere in the relatively recent past we and the apes share an ancestor, he said, a creature with both human and simian features. But where was this so-called missing link? The rocks were silent.

And so the search was on.

By happy chance, my walk along the prime meridian takes me to the Sussex village of Piltdown, not much more than a country crossroads but a place that played a famous—infamous!—role in the search for human ancestors. I stop for a pint at the Piltdown

Man pub, from whose sign a fossil skull looks down with a glittering eyeball, arched brow, and mischievous smirk. (See figure 4-1.) Before 1912 the pub was known as the Lamb. Then an amateur geologist and archaeologist named Charles Dawson made an extraordinary discovery while digging in a nearby gravel pit: a fragment of human skull. Dawson enlisted the aid of Arthur Smith Woodward, keeper of geology at the British Museum's natural history division (now the Natural History Museum). Together with a young French Jesuit paleontologist, Pierre Teilhard de Chardin, the pit was scoured for additional relics. More fragments of the skull turned up—they appeared decidedly human—and an apelike jawbone. Other animal bones and teeth and

FIGURE 4-1. The Piltdown Man pub, Piltdown.

primitive hand axes were found nearby, and a tool made from an elephant's femur that looked—to the delight of the English press when the discoveries were revealed—like an early cricket bat. The gravel beds in which the fossils were found are at least one hundred thousand years old, perhaps as old as 1 million years. When Dawson and Woodward announced their discoveries to the Geological Society, the news created a sensation. Not only had the long-sought missing link been found, but he was English!

A humanlike skull and an apelike jaw. Piltdown Man, as he came to be known (officially *Eoanthropus dawsoni,* Dawson's dawn man), was exactly the bridge between monkey and man that Darwin had anticipated and Wilberforce disparaged—the missing link. Who could now deny humankind's ancestral affinity to the apes? For the British champions of human evolution, the Piltdown skull was almost too good to be true. And, of course, it *was* too good to be true. From the very first there were doubters, from virtually every quarter of the scientific establishment except the British Museum itself. American and French paleontologists, especially, wondered aloud whether the skull and jawbone belonged to the same individual, and pointed out the remarkable similarity between the purported human-ancestor jawbone and that of a modern-day chimpanzee. Even a professor at King's College in London thought the jawbone might be that of a chimp. But Woodward, Dawson, and their supporters stuck to their guns, releasing salvos of pomp and bluster at the doubters, and poor Piltdown Man—with its human cranium and chimp jaw—staggered through the anthropology textbooks for forty years, although it became increasingly difficult to find a place for him among other hominid (human ancestor) fossils being discovered in Africa and the Far East. Eventually, more and more English paleontologists added their voices to the swelling chorus of skepticism.

Finally, in the 1950s, Joe Weiner, a South African physiologist who held a readership in anthropology at Oxford University, dared to say aloud the word that others had only whispered: *fraud.* A bit of serious investigation and suddenly the truth was out. The human skull fragments were relatively modern. The jawbone was that of an orangutan. The bones had been doctored, even painted, to make them look appropriately ancient and as if they belonged together. Once again Piltdown Man became an international sensation: Distinguished scientists had been victims of a hoax; Britain's premier scientific museum had been duped. The famous missing link was no link at all, and those who held firm to a special human creation gloated gleefully.

Who perpetrated the fraud? Dawson was a natural suspect. He died in 1916, not long after Piltdown Man was loosed upon the world. Some people pointed a finger of guilt at Teilhard de Chardin, who died at age seventy-four just a few years after the exposé; his motive, presumably, was to pull a Gallic joke on his English colleagues, a joke that spun wildly out of control. Hugh Miles, Joe Weiner's grandson, writing in the (London) *Sunday Times Magazine* in 2003, makes a convincing case against Charles Chatwin, a young staff member at the Natural History Museum who chafed under the leadership of his deeply unpopular and dictatorial boss, Arthur Smith Woodward. Chatwin's motive was to embarrass the pompous and gullible Woodward, says Miles. He enlisted the aid of another young staffer, Martin Hinton, and together with the witting or unwitting help of Dawson, they initiated a prank that turned into a full-scale hoax.

As I write, the Natural History Museum is exhibiting the Piltdown relics for the first time since they were tucked embarrassingly away in the basement in the 1950s. In retrospect, the entire Piltdown affair seems like a comic opera, but at the time of the "discovery," and again when the fraud was exposed, much seemed to be at stake. In the story of human origins, Piltdown was a side track that led nowhere. It was scientists who perpetrated the hoax, and scientists who exposed it. If there are lessons to be learned from the story, they are twofold. First, not even scientists are immune to the human tendency to believe what we want to believe. And second, this tendency is all the more reason to found our beliefs on empirical evidence, systematized skepticism, peer review, and all of the other epistemological apparatus of the so-called scientific method.

Few sciences have generated such vigorous controversies as paleoanthropology, the science of human origins. This is partly due to the fragmentary nature of the evidence, partly due to the strong personalities of some of the scientists involved, and partly because the whole question of human origins has been laden with emotional baggage at least since the time of Darwin. Nevertheless, a broad consensus among scientists has emerged concerning human ancestry, and the story told by the fossil bones has been generally confirmed by molecular biology. In recent years it has become possible to quantitatively compare the DNA and proteins of humans and other animals. For example, modern humans share 99 percent of their genes with chimpanzees, who appear to be our closest relatives among the primates. This does not mean, of course, that we are descended from chimps, but rather that we have a common ancestor with chimps somewhere deep in our past, probably about 6 or 7 million years ago in East Africa.

It is now manifestly clear that Sussex, England, is among the *least* likely places to look for "missing links." For one thing, the strata there are either too young or too old to be contemporaneous with early hominids. The oldest known human skull from England, the Swanscombe skull, dates from only about four hundred thousand years ago. It can be seen in the Museum of London (a different institution from the Natural History Museum), as part of an exhibit of the earliest human presence in the London Basin. The skull was found not far from my prime meridian walk, in gravel beds by the side of the River Thames, in three fragments, on three separate occasions between 1935 and 1955. It now reposes in a plexiglass case, glistening like bronze, by every test authentic. The skull belongs to a female in her early twenties, whose brain capacity lies well within the range of modern humans. Although her bones are old enough to make her species ancestral to *Homo sapiens,* she is not after all so different from ourselves. Carefully crafted flint hand axes from the same period speak eloquently of sophisticated mental activity.

The Swanscombe woman is a recent human ancestor compared, say, to Lucy, the fairly complete skeleton of a female hominid who lived in East Africa about 3.2 million years ago. Today, Lucy's bones lie in the National Museum of Ethiopia, where she is tagged with the more prosaic appellation A. L. 288-1. But her evocative image is available to us in a splendid book, *From Lucy to Language,* which her discoverer Donald Johanson coauthored with the science writer Blake Edgar, illustrated by the fossil photographer David Brill. The first half of the book is a review of the goals, methods, and discoveries of paleoanthropology, handsomely illustrated with Brill's photographs. The second half of the book, called "Encountering the Evidence," is a sort of family album, a gorgeously presented survey of the most important hominid fossils, with each skull, jaw, or collection of bone fragments glistening in a rich bronze patina against a black background.

Lucy is the star of the album—forty-seven bones, about a fourth of a complete skeleton, enough to let her come alive on the page. She is a fine example of *Australopithecus afarensis,* now commonly considered ancestral to all—or almost all—later hominid species. She was a little over three feet tall, with long arms. That she walked erect is confirmed by a fine trackway of *A. afarensis* footprints discovered at Laetoli, Tanzania, in 1976. Two individuals walked side by side through fresh volcanic ash, perhaps a parent and child; at one point, they seem to have paused and turned to look toward the west. One of their footprints is reproduced in the Brill album; the impression reveals a strong heel strike, the longitudinal arch and ball of the foot, and a deep indentation of the big toe. Across the millions of years, the Laetoli print invites us to place our own bare foot into the impression, and to feel the bond of ancestry that links us to Lucy's kind.

As one turns the pages of the Lucy family album, we confront a series of skulls or partial skulls that brings us through the many-branched family tree of Lucy's descendants. Some of these individuals belonged to lineages that became extinct. Others led ultimately to modern humans by pathways that are not yet entirely clear. Although the precise genealogy of *Homo sapiens* is still hotly contested, the general sweep of development is clear from the photographs: As we move forward in time, the skulls become ever more unmistakably modern, more closely resembling our own. The evidence for our family tree is sparse, but year by year the evidence grows more voluminous, and year by year the essential story it tells becomes more irresistibly clear. Darwin's prescient guess was right: We are distant cousins of the chimps. Our shared original home was apparently the volcanic grasslands of East Africa.

But there is another way to look at the fossils in the family album. Each of these gleaming skulls was an individual, with a unique identity. Behind the gaping eye sockets was a dawning self-awareness, although we may never know the precise moment when lips first formed the words "I love you," or "I am afraid," or "Look at the beauty of the night." We turn the pages of the album, forward, backward, searching the fossil fragments for the true beginning of humankind. Did it occur when crude tools first begin showing up with the bones? Or when rock paintings and carved figurines make their appearance? Or with evidence for deliberate burial with funerary goods? Or was it earlier, several millions of years ago, when an *A. afarensis* female paused as she crossed a field of warm volcanic ash, gripping her offspring's hand more firmly as she looked away to the west, to the danger of an erupting volcano. That gentle, protective squeeze of a child's hand at the dawn of time may be the moment in our family history when we became recognizably different from every other creature on planet Earth, destined for a life of conscious thought, moral responsibility, and cosmic wonder.

Pity the poor English fossil hunters. It was the Englishman Charles Darwin who most forcefully argued the antiquity of human ancestry in his 1871 *The Descent of Man,* but the English strata were stubbornly ungenerous in yielding evidence of humankind's deep past. Already in 1857 a beetle-browed skull cap had been found in a cave in Germany's Neander Valley, and soon other "cave man" relics began turning up all over continental Europe. These so-called Neanderthals were too humanlike to be "missing links," yet they were anatomically different from modern humans. The general consensus of the time was that they were a rather more brutish sort of human—hulking, hairy, thick-necked—ancestral to our more slender and refined selves. To be sure, British anthropologists found tools of stone and bone that suggested an early human or hominid presence in Britain, but the creatures who made the utensils remained frustratingly elusive. Only near the end of the nineteenth century did fossils that were claimed to be Neanderthal turn up in a gravel pit at Galley Hill, near the Swanscombe site on the banks of the Thames, but their antiquity was uncertain. Were the bones laid down by the river with the gravel, or were they part of a recent burial? ("Galley Hill Man" was subsequently shown to be a modern woman, perhaps a victim of the gallows from which the hill takes its name.) No wonder Woodward and his fossil-starved British colleagues latched onto late-coming Piltdown Man with such uncritical delight.

If one does an Internet search for "Piltdown Man" or "Galley Hill Man," most of the hits will be sites maintained by Christian individuals or groups who hold firm to the literal Genesis story of special creation. Nothing delights creationists more than evolutionists making fools of themselves. If scientists could be wrong about Piltdown Man and Galley Hill Man, they crow, then they could be wrong about other things too. But of course scientists are human and as prone to folly as the rest of us. What the Piltdown and Galley Hill stories demonstrate, if anything, is the willingness of scientists to change their minds when a preponderance of evidence goes against them, a characteristic decidedly lacking among those who believe that truth has been revealed once and for all by divine communication. The reason the scientific community universally supports an evolutionary view of human origins has very little to do with the authenticity of any single piece of evidence, but with the way the overall picture hangs together, including not just anthropological evidence but also genetics, chemistry, physics, geology, paleoclimatology, and all the rest.

The story of human origins, like the other stories I have followed on my walk along the meridian, takes us one more step away from the place where our species' intellectual history began, and where each of us begins as individuals: at the presumed center of space and time. Even when we recognize the antiquity of our species and our kinship to the beasts, the tendency to see ourselves as special can warp our judgment—as is nicely illustrated by the story of Neanderthals.

My own introduction to Neanderthals came with a book in my parents' library, H. G. Wells's *The Outline of History,* published in the 1920s. An illustration in the book shows a Neanderthal male, with a face that is sour and simian, and dull, squinty eyes. (See figure 4-2.) "It's thick skull imprisoned its brain, and to the end it was low-browed and brutish," wrote Wells, reflecting the prevailing opinion of the time. Neanderthals were then well-known from many skeletal remains from Europe and western Asia. It was clear they had lived in those places for several hundred thousand years, during the height of the ice ages, only to disappear from the fossil record about thirty thousand years ago. According to the standard story of Wells's generation, the "hairy," "ugly," "dim-witted" Neanderthals were replaced—and rightly so, presumably—by the nimble, bright-eyed, intelligent Cro-Magnons (see figure 4-3); that is, by our own superior race.

But, of course, bits of skull and bone reveal nothing about personality, intellect, language, or culture, and very little about what might have caused the Neanderthals' extinction. Wells based his appraisal as much upon an assumption of *Homo sapiens* superiority as upon solid archaeological evidence. He spoke for science, but his voice was loaded with disparaging inference. Note,

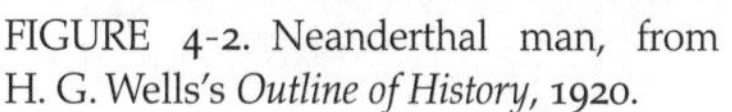
FIGURE 4-2. Neanderthal man, from H. G. Wells's *Outline of History*, 1920.

FIGURE 4-3. Cro-Magnon man, from H. G. Wells's *Outline of History*, 1920.

especially, his use of the impersonal pronoun *it* when referring to Neanderthals.

Recent evidence from the field tells a rather different story.

Recovery of mitochondrial DNA from a Neanderthal skeleton several years ago suggests that Neanderthals and modern humans diverged from a common stock at least half a million years ago, probably in Africa, then evolved along parallel lines. Ancestors of the Neanderthals eventually arrived in Europe and western Asia, where they thrived near the margins of ice age glaciers. Apparently, they made stone tools, clothing and shelter, used fire, decorated their bodies with ornaments, and at least occasionally buried their dead. There is circumstantial evidence that they cared for the aged and handicapped. Their brains were as capacious as our own.

Then, around forty thousand years ago, Neanderthal territories were invaded by Cro-Magnons, our immediate *Homo sapiens* ancestors. For thousands of years the two branches of the human family lived side by side. There is no convincing evidence of interbreeding; they may have been separate species, unable to produce offspring. For one reason or another, Neanderthals were slowly driven to oblivion; their last redoubt seems to have been the southern part of the Iberian Peninsula. It may be that Neanderthals were no match for their more technologically advanced neighbors, in much the same way that the indigenous people of the Bahamas, the Lucayans, were eliminated from the face of the Earth by the arrival of more technologically advanced Europeans. Certainly, the Neanderthal extinction is one of the large dramas of human history—a devastating instance of deliberate or accidental intraspecies genocide—and certainly the most momentous loss of biodiversity ever caused by *Homo sapiens.*

History is written by the winners, as Ian Tattersall, curator of the Department of Anthropology at the American Museum of Natural History in New York, reminds us, and the story of Neanderthals is written by Cro-Magnon descendants for an audience of Cro-Magnon descendants. Neanderthals were losers, and there is no more irrevocable way of losing than extinction. When Wells presented Neanderthals as "ugly" and "dim-witted," he was merely doing what winners have often done. Virtually all people in all times have considered those outside of tribe or kin as somehow inferior.

History may be written by the winners, but winners can change what they write. An update on Wells's reconstruction of a Neanderthal face can be found in Tattersall's *The Last Neanderthal* (1995). This new fellow has wide, curious eyes and a slightly bemused expression. He could be anyone's kindly grandfather. Put him in a plaid shirt and pair of overalls, and he would not attract all that much attention as a fellow passenger on a crosstown bus. Tattersall's sympathetic view of Neanderthals is typical of the current generation of anthropologists. His account is laden with inference, as all anthropology must be, but he goes out of his way to avoid prejudicial language and puts the most generous spin on the evidence. In Tattersall's account the brutish subhumans of Wells's story give way to a people who are the gentle, intelligent victims of Cro-Magnon violence.

This change of opinion is at least partly driven by a sea change in the way we value alien cultures—the same transformation that led to a reevaluation of the role of Columbus and his successors from religious saviors of savages to exterminators of a less technologically endowed people. The Cro-Magnon invaders of Europe probably never questioned their right, even obligation, to kill the indigenous—and alien—Neanderthal inhabitants of those lands. It was kill or be killed, and Cro-Magnons presumably possessed the better technology. Evolution is a story of never-look-back competition, red in tooth and claw. Humans, alone of all species, do sometimes look back. We cultivate a sense of history. We ask ethical questions about past actions.

Of course, it is hardly fair to impose contemporary moral standards upon our ancestors, especially those of the very distant past, but in asking ethical questions about human history, and in revising answers, we redefine ourselves. I once heard the anthropologist Margaret Mead say that the progress of civilization is the ever-widening circle of those whom we do not kill. Perhaps we have at last become civilized enough to recognize the injustice of exterminating a people who may have been a separate species, but who were nevertheless an intelligent, cultured part of the human family. Alas, our enlightenment comes too late for the Neanderthals.

5

COSMIC TIME

My walk along the prime meridian brought me inevitably to the Royal Observatory at Greenwich, founded by King Charles II in 1675. Charles had come to the throne fifteen years earlier, after England's troubled experiment with nonroyal rule under the violent and puritanical Oliver Cromwell. With the restored monarchy, drama, arts, and literature flourished again. Science, in the sense that we know it today, was born as an organized social activity with its own protocols for the international transmission and verification of knowledge. Anyone who has studied physics in secondary school or college has heard of Hooke's law of elasticity, Boyle's law of gases, and Newton's laws of motion. Many more people are familiar with Pepys's diaries, Wren's churches, and Halley's comet. All of these eponymous gentlemen glittered in the intellectual firmament of late-seventeenth-century London, a city with one fifth the population of present-day Toronto. They knew each other, fed off one another's curiosity, and together established the Royal Society, the first bona fide scientific organization. They can fairly be called the first moderns, and that is exactly what they understood themselves to be. To my mind they were taking up where the Alexandrians left off nearly two thousand years before.

Standing head and shoulders above all of the great scientific thinkers of Restoration England was Isaac Newton, a man of prodigious intellect and inscrutable personality. Here's how his biographer James Gleick describes him: "He was born into a world of darkness, obscurity, and magic; led a strangely pure and obsessive life, lacking parents, lovers, and friends; quarreled bitterly with great men who crossed his path; veered at least once to the brink of madness; cloaked his work in secrecy; and yet discovered more of the essential core of human knowledge than anyone before or after." The last clause of that sentence is a formidable claim for any biographer to make of his subject, especially of a man whose abstruse great book, *On the Mathematical Principles of Natural Philosophy* (or simply, from its Latin title, *Principia*), had few comprehending readers then or now. Can the claim be justified?

Consider Newton's achievement in the years 1665–66, while still a student at Trinity College, Cambridge. Plague had reached England; in London one of every six people would die, and soon the contagion spread to outlying towns. The colleges of Cambridge closed down, and students dispersed to the countryside; Newton returned to his home in Woolsthorpe, Lincolnshire. (The house is now in the care of the National Trust, a place of scientific pilgrimage.) Solitary and self-taught, he embarked upon a spree of mathematical discovery that seems almost superhuman. He developed the theory of infinite series and showed that it was possible to treat the infinite and the infinitesimal with mathematical rigor. He refined concepts of space, time, inertia, force, momentum, and acceleration, and formulated laws of mechanical motion. He invented the theory of universal gravitation and applied it to celestial and terrestrial motions, showing that Kepler's three laws of planetary motion followed inevitably. To facilitate his calculations, he devised what is now called the differential and integral calculus. All of this before he was twenty-four years old.

Astonishingly, Newton kept most of it to himself. At age

twenty-seven he became Lucasian Professor of Mathematics at Cambridge and occasionally sent dribbles of what he had invented for consideration by the Royal Society in London. The irascible genius Robert Hooke, in particular, was quick to give Newton a hard time, frequently claiming priority for himself. Newton's response was to become even more secretive. In solitude he continued his mathematical and physical studies but also embraced alchemy and esoteric biblical scholarship. He felt himself to be seeking ancient knowledge that had been lost or hidden in the dark centuries of the recent past. Scientists today tend to be embarrassed by Newton's religious and alchemical studies, but Newton was looking for deeper, unifying truths than what he considered to be the superficial speculations of the secular empiricists of London. He wanted to read the mind of God.

Enough of Newton's ideas leaked out to make the London savants aware that the solitary professor at Cambridge sat atop a trove of new learning. Eventually, Edmond Halley, later of comet fame, persuaded Newton to write his discoveries down, and then published the book at his own expense. Suddenly Newton was wildly famous, extolled by scientists and poets alike. Alexander Pope's couplet summarizes the adulation: "Nature and Nature's laws lay hid in night; God said, *Let Newton be!* and All was *Light.*" Newton left Cambridge for London, leaving not a friend behind, and took up posts as warden of the Mint and, eventually, when Hooke died, president of the Royal Society.

If there is a single work of human genius that separates medieval and modern times, it is the *Principia.* In it Newton demonstrated that the motions of Sun, Moon, Earth, planets, the moons of planets, comets, tides, cannonballs, and falling apples can all be deduced with mathematical exactitude from a handful of elegantly simple laws of nature. The Newtonian world is not an arena of spirits and divinities acting arbitrarily; rather, it is a great clockwork, whose every tick has been determined since the dawn of time by immutable mechanical laws.

When I worked through the astronomical theories of Ptolemy, Copernicus, Tycho, and Kepler during my 1968–69 sabbatical sojourn in London, I did not need to include the theory of Newton. Indeed, I had already done so—as an undergraduate physics student. *Every* undergraduate physics student does it: Fold together Newton's law of universal gravitation (the force of gravity between any two objects is proportional to the product of their masses and inversely proportional to the square of the distance between them) and Newton's second law of motion (force equals mass times acceleration), and solve the resulting second-order differential equation. Out pops Kepler's laws of planetary motion. Out pops an exact description of the orbit of Mars in 1968–69 and every other year past and future. Out pops the entire solar system, the precise detailed motions of planets, moons, asteroids, and comets. Out pops, yes, even planets that were unknown to the ancients, Uranus and Neptune, which revealed their presence as tiny perturbations in the motions of the known planets even before they were observed in the sky. All of it—all of it is contained majestically and almost magically within a single equation. Today, when NASA scientists send a space probe to Mars or Saturn's moon Titan, for example, they use nothing other than Newton's laws of gravity and motion to guide it upon its course. This was Newton's astonishing achievement.

The trajectory of Newton's life never took him far from the prime meridian—from his birthplace at Woolsthorpe, in Lincolnshire, to his final posting at the Royal Mint in London. The most creative part of his life, of course, is associated with Trinity College, Cambridge. Step through the gate of Trinity today, and you are carried back to the seventeenth century; not much has changed since Newton's time that is visible to the eye. A short stroll across the enclosed courtyard brings you to the antechamber of the college chapel. Here are statues of the college's leading scholars, dominated in the place of honor by an impressive likeness of Newton, with the inscription *Qui genus humanum ingenio superavit,* which translates (very) loosely from the Latin as "Nobody smarter." Newton's book *On the Mathematical Principles of Natural Philosophy* (the *Principia*) has often been called the most important philosophical work ever published. It is also probably the least-read philosophical work. Few people then or now have the patience or the skill to wade through the dense mathematics. Although Newton invented the integral and differential calculus to solve his problems, he wrote the book in a more abtruse geometric language, rightly, I suppose, because he saw the futility of writing a book in a language that no one understood but himself. In fact, the differential and integral calculus (invented at the same time independently by the German mathematician Gottfried Leibniz) quickly became the language of physics precisely because it was so perfectly suited for solving the kinds of problems Newton considered. The calculus is a purpose-made language for treating infinitesimal, indistinguishable instants of time and infinitesimal, indistinguishable intervals of space. The Greek geometric astronomy of centered, uniformly turning circles had had its day. With Newton, the universe was no longer conceived of as a cosmic egg with humans at the center, presided over by a Creator who acts at will in his creation. In Newton's physics space is infinite and time eternal, both are uniform and homogeneous, and God, if he enters into the story at all, is confined to some vague creative activity as the instigator of the immutable laws of nature.

With Newton we have come about as far as we can go from the sacred space and time of our remote ancestors, or even from the space and time of believing Christians of Newton's own time. "All the world's a stage," wrote Shakespeare, and he meant it literally. The space and time of Shakespeare's cosmos—like that of Newton's contemporary John Milton—was *dramatic* space and time, a divinely created venue for the human drama that began with the creation and fall of Adam and Eve, reached its central dramatic moment with the death and resurrection of the Son of God, and would ring down the curtain with the Apocalypse and Second Coming. In place of this sacralized, human-centered cosmos Newton offered something stark and pure. "Absolute, true, mathematical time, of itself, and from its own nature, flows equably without relation to anything external," he wrote. And further, "absolute space, in its own nature, without relation to anything external, remains always similar and immovable." In this yawning, perhaps infinite emptiness, which exists without relation to anything external to itself, the planets, moons, and stars have their respective motions like motes of dust in the vast empty space of a cathedral. The "mathematical principles of natural philosophy" that apply throughout Newton's cosmos make no mention of human history. His physics is as cold and stately as the marble statue of the man himself that stands in the antechamber of Trinity College chapel.

Newton's laws of force and motion describe with stunning exactitude the courses of comets and the fall of an apple from a tree, but they provide no insight into the mind of Newton. He lived in a strange and convoluted psychological universe, as different as can be imagined from the orderly mechanical universe described in the *Principia.* In friendless solitude he pursued his mathematical and physical studies, but (as we have seen) he also dabbled in alchemy and esoteric biblical scholarship. He was such a private man that we will never know what went on inside his head. He perhaps saw more deeply into the nature of the cosmos than any man before or since (with the possible exception of Albert Einstein), and yet he was able to write, "I seem to have been only like a boy playing on the seashore, and diverting myself in now and then finding a smoother pebble or a prettier shell than ordinary, whilst the great ocean of truth lay all undiscovered before me." He lived on a cusp of history, which we now call the divide between the pre-Newtonian and post-Newtonian worlds. The publication of the *Principia* can be taken as the first moment of modernity.

What is it like to live on a cusp of history? The best introduction I know of to the intellectual milieu of Restoration London is the diary of Newton's contemporary Samuel Pepys. Like many other students, I read bits of the diary as a high school student, accounts of the Great London Fire of 1666 and the visitation of the plague to London in 1665. What we read was severely expurgated, as I discovered relatively late in life by reading the entire unedited diary, almost ten years' worth of entries and what seems like a zillion words. Pepys is as uninhibited on the page as Newton was inscrutable.

Most of what Pepys committed to paper is a record of his pursuit of pleasure: food, drink, music, drama, and women. During the years of the diary, 1660–69, Newton was at Cambridge formulating his world-transforming theories in monastic solitude.

Londoners were enjoying a gay debauch, following the example of their fun-loving monarch, Charles II. The king's pleasures were unabashedly public, his mistresses openly acknowledged. Pepys was somewhat more discreet; he kept his amorous adventures secret from his wife, at least until she caught him in flagrante delicto with her "lady." What makes Pepys's diary such compelling reading is the combination of private and public history. He was a man about town who hobnobbed with royalty, nobility, intellectuals, artists, military men, clergymen, and bishops, as well as tarts, boatmen, hackney drivers, and tavern keepers. He might move directly from an audience with the king at Whitehall Palace to a saucy performance by Nell Gwyn at the Duke's theater. His diary is an portrait of his age in all of its gaudy detail and subtle nuance.

After England's grim interlude with Cromwell and puritan Roundheads, Londoners can perhaps be excused for a bit of carrying on. The 1660s were not only defined by pleasure, however. The city was also home for many of the ingenious "savants" and "virtuosos" (as they called themselves) who created science as we know it today. New ideas were in the air, new ways of wresting knowledge from nature. The Royal Society was a glittering assembly of genius. Pepys was a government bureaucrat, a senior naval administrator, not a scientist. Nevertheless he was swept along by the excitement. He purchased every new science book that came off the press, and struggled to understand it. He bought a microscope and a telescope and almost every other clever device that defined the new experimental age. And he cultivated the friendship of scientists. In 1665 he was elected to membership in the Royal Society; later, he would become the society's president.

Pepys's interest in science was partly intellectual curiosity and partly fashionable fad. When he acquired a "perspective glass"—

an early form of binoculars—the first place he took it was to church, where from a pew in a gallery he "had the pleasure of seeing and gazing at a great many fine women." A new age was being born, defined by a conviction that the world is ruled by natural laws that can be discovered by human reason. But the old ways lingered. One moment Pepys might be observing an experiment on blood transfusion, and the next he is at Charing Cross to see some perceived enemy of the realm drawn and quartered as a kind of spectator sport. One moment he listens to Robert Hooke speculate that comets are periodic objects that obey exact mechanical laws, and the next he worries that the year 1666 is characterized by "666," the number of the Beast of the Apocalypse. In his diary entry for January 21, 1665, Pepys attributes his good health to the influence of his new rabbit's foot, a lucky charm. Then he sits up late reading Hooke's *Micrographia,* a book that records some of the first scientific observations with a microscope. Among the famous illustrations in that book is one of a flea, sketched by Hooke with every bristle, crease, and scale, and published in the very year that plague ravaged London, killing thousands and virtually shutting down the city. We now know that plague is caused by flea-borne bacteria.

In other words, Pepys was a man of his time, and his time was the Age of Reason half born. We no longer participate in public executions, carry rabbit's feet, worry about apocalypse, fear comets, or die of plague. Or at least most of us don't in the developed Western world, where, since Newton, science has held sway. And the reason, of course, is *reason*—as embodied in the scientific way of knowing created by the contemporaries of Pepys. It is perhaps not a coincidence that science got its institutional start during the reign of the debauched Charles II. For all of his sorry human failings, the monarch encouraged religious tolerance, artistic vivacity, and intellectual freedom. If history is a guide, science thrives best in societies that are secular, democratic, and free.

When King Charles II issued a Royal Warrant establishing a national observatory on the highest ground of the king's parkland at Greenwich, on June 22, 1675, he was not motivated by disinterested curiosity about the natural world. Charles had a navy to watch out for, far-flung colonies, and a world to plunder. His kingdom prospered or foundered with trade. Among masters of English ships—as indeed among all sailors of all nations—one problem reigned supreme: the determination of longitude.

As we saw in an earlier chapter, it is easy enough to know one's latitude at sea; that is, degrees north or south of the equator. One need only measure the angular elevation of a celestial object—Sun or star—above the horizon as it crosses the local meridian. But knowing where one is east or west is another matter. Sailors were forced to rely on "dead reckoning," keeping track of their east-west progress by guessing the speed of the ship. The method was unreliable at best, and many a ship came to ruin by finding itself unexpectedly dashed against an inhospitable shore.

One answer to the problem of longitude, as everyone knew, was having access to a clock at sea that accurately kept the time of one's home port. The Earth turns on its axis fifteen degrees per hour. Therefore, if a shipboard clock keeping Greenwich time, for example, reads one P.M. when the Sun reaches its highest point *in the ship's sky* (local noon), then one has sailed fifteen degrees to the west of Greenwich. If the shipboard clock keeping Greenwich time reads midnight when the Sun is highest in the sky, then one has sailed halfway around the world from Greenwich. Time equals distance. Four minutes' difference between shipboard time and home-port time equals one degree of longitude.

One degree of longitude can be as much as seventy miles at sea (depending on how far one is north or south of the equator), and seventy miles can mean the difference between safe haven and disaster. No mechanical clock of the late seventeenth century could keep time accurately enough on a long voyage to be of much use for navigation. If on a voyage of a month a clock lost or gained even a minute—a hopelessly unattainable standard of accuracy in the seventeenth century—this could translate to a dozen miles or more at sea.

But perhaps an accurate "clock" could be found in the night sky. The moons of Jupiter observed with a telescope make a fine celestial clock, as Galileo realized, but telescopes are of little use on a rolling ship. The movement of the Moon against the background of fixed stars held promise as a kind of celestial clock, but first accurate tables of predicted lunar positions were required, and for this an observatory was necessary. And so at the king's bidding Christopher Wren designed the handsome building that still stands gloriously intact on the hill in Greenwich Park, and John Flamsteed was appointed first astronomer royal.

A typical visitor to Greenwich today catches a first glimpse of the Royal Observatory through the colonnaded walkway that connects the National Maritime Museum to the seventeenth-century Queen's House designed by Inigo Jones for the wife of James I. The observatory crowns the hilltop across a rising swath of green. (See figure 5-1.) Atop the observatory is the large orange Time Ball that rises each day along its staff at five minutes before one P.M., and falls exactly on the hour, as it has done each day since 1833 when it was set up as a way for departing mariners

FIGURE 5-1. Royal Observatory, Greenwich.

along the Thames to set their shipboard clocks. (The astronomers chose one o'clock as the signal time because at noon they were busy making observations of the Sun as it crossed the local meridian.) Of course, today maritime traffic on the Thames takes its time cue from a radio signal, and the rising and dropping of the Time Ball is a mere historic curiosity, but as one o'clock approaches nearly everyone in Greenwich Park pauses and turns toward the observatory to watch an event that reaches across the centuries to connect us with humankind's long search for standard time. I have on several occasions checked my own watch against the dropping of the Greenwich Time Ball.

My most recent visit to the Royal Observatory was on my

meridian walk, which of course brought me directly to the observatory. I arrived early, just as the observatory was opening to the public. My goal was to have some minutes alone in Flamsteed House, the original building designed by Wren, and in particular in the Great Star Room, or Octagon Room as it is called today. The room encompasses the entire second story of Flamsteed House. Its tall windows give access to the sky on every side but two of the octagon. The room looks essentially today as it did in the time of Flamsteed, Wren, Hooke, Halley, and Newton. (See figure 5-2.) Two clocks set into the wall, with thirteen-foot pendulums swinging behind the wainscoting, tick for me as they ticked for Flamsteed. (The present clocks are copies of the originals; Flamsteed took the originals with him when he retired as astronomer royal; one of the original movements has been recently restored to the observatory and is on display in the Octagon Room, the other is at the British Museum.) These were among the most accurate clocks in the world in 1676, when the astronomer royal purchased them for the observatory from Thomas Tompion, the best of English clockmakers. *Tick* is not exactly the right word; in the solitude and silence of the early morning, the sound of the clocks in the Octagon Room is really more of a dull *tunk-tunk.*

Sunlight streamed through the tall windows. In an hour the room would be swarming with schoolchildren and tourists. I stood quietly near the Tompion clock and tried to imagine the young Galileo, as a student at Pisa, sitting in the shadowy cathedral of that city, watching a hanging lamp swing slowly back and forth. According to a later account of his student and assistant Vincenzo Viviani, Galileo observed that the time for a single oscillation (the period) of the lamp remained constant as the lamp's oscillations diminished in amplitude; that is, the lamp required exactly as much time to swing through a small arc as through

FIGURE 5-2. The Octagon Room, Royal Observatory, Greenwich.

a large arc. If the event occurred as described by Viviani—another early biographer asserts that the lamp of the story did not exist in the cathedral when Galileo was at Pisa—one can only suppose that Galileo used his pulse to ascertain the isochronism (constant period) of the swinging lamp. Galileo may also have observed the isochronism of a pendulum while he was being instructed in music by his musician father. However it happened, it was an important discovery, and the use of a pendulum as a timekeeper would lead at the hands of the Dutchman Christian Huygens to the first mechanical clocks of sufficient accuracy for scientific work. Galileo himself experimented with pendulum clocks all of his life, and even considered the possibility of using a shipboard pendulum clock for determining longitude, but never succeeded in producing a clock of the requisite accuracy. The clocks Flamsteed acquired from Tompion used a pendulum as the driving mechanism and were among the most accurate clocks of their time. Although they served Flamsteed well for his astronomical work, they were not suitable for the rigors of a voyage at sea.

Consider for a moment Galileo in the cathedral at Pisa. (Don't worry if the event happened as described by Viviani.) Perhaps a minor official of the cathedral came by to light the hanging lamp, and set it swinging in the process. As the lamp slowly returned to rest, the young student watched, and, from his innate sense of the passage of time, intuited the isochronism. Alone in the cathedral, he may have then given the lamp a shove himself, setting it moving again in a long languorous arc. With his pulse he timed the swing, taking an average over a number of cycles. As the lamp's oscillation dampened to an almost imperceptible amplitude, he timed the cycle again. The same! Immediately Galileo recognized that the thing he had discovered might have practical application to timekeeping.

But how could he be sure that the pendulum's motion was truly isochronous? After all, perhaps his pulse rate was not constant. Perhaps his pulse rate speeded up just enough to make the pendulum *seem* isochronous when in fact the lamp was taking less time to move through the smaller arc. Galileo was surely aware from his medical studies that pulse rates vary, between individuals and even for the same individual. In fact, he would later try to adapt a pendulum mechanism for measuring pulse rates, turning his discovery around. The point here is that any measurement of time requires a timekeeper. The accuracy of the timekeeper can only be determined by comparing it to another timekeeper, presumed to be more reliable. So which is the more reliable timekeeper, the swinging lamp or the pulse? It all seems terribly circular. Flamsteed wanted Tompion's clocks so that he might time celestial motions, but the accuracy of Tompion's clocks could only be ascertained by comparing their rate of ticking to the very celestial motions Flamsteed wanted to measure. Is there any way out of the circularity? Newton imagined a kind of time that flows equably without reference to any timekeeping device or the motion of any celestial body. But how is this abstract Newtonian time to be perceived except through the agency of some pulsing timekeeper?

Let us suppose that the lamp swinging in the cathedral at Pisa is *not* isochronous; that is, it takes more time to swing through a large arc than a small arc. But large and small arcs correspond to exactly the same number of pulse beats, as counted by Galileo. This means that Galileo's pulse rate must have speeded up *by exactly the amount necessary to make the swinging lamp appear isochronous.* This would be a highly unlikely coincidence, especially if Galileo repeated his observations several times, perhaps on different days, and always with the same result. So it is the *apparent* isochronism of the lamp that Galileo takes as the guarantor of the lamp's *absolute* isochronism. There is an elegance about the very notion of isochronism, and Galileo was guided all his life by the idea that nature's laws are simple and elegant. And surely, he must have imagined, a swinging pendulum, once its isochronism was discovered, is a more trustworthy keeper of nature's time than a human pulse rate, which was known to be susceptible to wild perturbations.

Likewise, Flamsteed bought *two* clocks for the Royal Observatory so that he could compare one against the other. I suppose if he could have afforded it, he would have had a dozen clocks all ticking in unison. The first thing he did was adjust the clocks so that they kept solar time. That is, their rate of running must be adjusted so that their dials recorded the passage of twenty-four hours between one passage of the Sun across the local meridian and the next. But it's the pendulum-and-pulse question all over again. How can Flamsteed be sure that the Earth turns on its axis at a steady rate? And the answer is the same. According to his measurements, the Earth's rotation rate compared to his clocks *appeared* to be constant. What were the chances that the clocks' rate of running would change by exactly an amount to make an inconstant Earth appear constant? In the end, Flamsteed assumed that his two Tompion clocks with the thirteen-foot pendulums (a two-second beat) were the closest thing he could get to Newton's absolute, universal, and utterly abstract time.

Today, we use clocks that take as their timekeeper not pendulums but the vibrations of atoms. The thirty-five-dollar watch on my wrist is controlled by a vibrating quartz crystal; it is far more accurate than the restored Tompion clocks that *tunk-tunk* in the Octagon Room of the Royal Observatory. When quartz crystal clocks were invented in the 1930s, it was discovered that the Earth's rotation is *not* exactly uniform; in fact, the rotation is slowing down, and each day is longer than the preceding day by an infinitesimal amount. But this assertion can only be made because we assume that atomic vibrations are likely to be more fundamentally isochronous than spinning planets, especially since we have theoretical reason to believe that gravitational forces between Earth and Moon can affect the spin rate of the Earth. Yes, there is an arbitrariness and circularity in all of our theories of the world; what makes one theory more reliable than another is the economy and elegance of the entire interlocking system of ideas.

On October 22, 1707, Royal Navy ships under the command of Admiral Sir Clowdisley Shovell sailed upon treacherous ledges of rock near the Isles of Scilly, off the southwestern tip of Cornwall. Four ships were lost, with two thousand lives, including that of the admiral. (His body was recovered from the sea and now lies in one of the most flamboyant tombs of Westminster Abbey in London.) It was a huge calamity, and although the causes of the disaster were many, it gave new urgency to the search for a reliable way to find longitude at sea.

In 1714 Parliament responded by establishing the Board of Longitude and offering a prize of twenty thousand pounds—a considerable sum of money in those days, more than a million dollars in present currency—to anyone who could provide a way to determine longitude at sea to within half a degree. At the latitude of the Isles of Scilly this corresponds to about twenty-seven miles. Since time equals distance, the terms of the prize meant devising a way for a clock to keep Greenwich time to within two minutes during a voyage of many weeks, specifically (according to the terms of the prize) a voyage across the Atlantic to a port in the West Indies. To have some idea of the challenge, consider that

Flamsteed's Tompion clocks went astray by as much as a minute every few days, and these were securely housed behind the woodwork of a solid building on a firm foundation. Even then Flamsteed complained about the effects of dust, lack of lubrication, movements of air, and changes in temperature on the running of his clocks. Aboard a tossing ship at sea, in wild weather, there was no hope that a clock such as Tompion's could keep the requisite time.

At the age of seventy-two, Newton was called upon to advise the parliamentary committee. He testified: "One [method of determining longitude] is by a Watch to keep time exactly. But, by reason of the motion of the Ship, the Variation of Heat and Cold, Wet and Dry, and the difference of Gravity in different Latitudes, such a watch hath not yet been made." And this, of course, is why the Royal Observatory had been established in the first place: to find some way of using the heavens as a timekeeper at sea. For the next century, one astronomer royal after the other knocked his head against the problem, with only limited success. The most promising astronomical solution to the problem of longitude was that of lunar distances: measuring the angular separation of Moon and Sun by day or Moon and stars by night, the Moon in effect playing the role of a moving hand on a celestial clockface. This task was made easier and more accurate by the invention of the handheld sextant, which had come into wide use by the middle of the eighteenth century. (Newton, Hooke, and Halley all had a role in the evolution of this instrument, although John Hadley in Britain and Thomas Godfrey in America are credited with the invention.) The Reverend Nevil Maskelyne, astronomer royal from 1765 to 1811, brought the lunar-distance method to a high degree of usefulness by publishing a nautical almanac that enabled sailors in far-flung corners of the globe to compare observed positions of the Moon to those Maskelyne had computed for Greenwich. The almanac of lunar distances played the role of a shipboard clock keeping Greenwich time.

Of course, the method of lunar distances requires clear skies and is of little use in storm-tossed seas, that is, in the very circumstances when a sailor most desperately needs to know his position. The solution to the problem of longitude, when it came, was by way—after all!—of a ticking machine you could hold in your hand—the marine chronometer—the invention of a mechanical genius, John Harrison, who lived, by delightful coincidence, at Barrow-on-Humber in Lincolnshire, near the place where the prime meridian leaves the British landmass northward on its henceforth over-sea course to the pole.

Harrison's story is that of a man who spent his entire life pursuing a single obsession: a clock that was immune to the vagaries of gravity and the elements, a clock that ticked in consonance with Newton's absolute time, a thing of brass and steel that was, insofar as was humanly attainable, a suitable heartbeat for the universe. Harrison produced four clocks for the Admiralty. The first satisfied the criteria of the competition on a voyage to Portugal and would eventually win for its maker half of the prize from a hesitant Board of Longitude, but it did not satisfy Harrison himself. The watchmaker spent decades improving his work, inventing devices to eliminate friction, compensate for changes in temperature, and regulate the force of the driving springs. Even then, having satisfied the requirements of the Board of Longitude, Harrison was forced to engage in a lifelong battle to obtain the promised prize against the machinations of those who acted from stinginess, stubbornness, or envy, not least of whom was Maskelyne, the astronomer royal, who was committed to the method of lunar distances and brooked no competition from "mechanics."

All four of Harrison's clocks are now on display in thick plexiglas cases in a museum room of the Royal Observatory. They are referred to collectively, simply, as the Harrisons.

The first of Harrison's clocks, H-1, completed in 1735 (with the help of his brother James), looks nothing like what we expect a clock to be. (See figure 5-3.) It is undeniably a thing of beauty, its shape eerily evocative of Flamsteed House itself. The clock's original

FIGURE 5-3. John Harrison's H-1, the first mechanical clock capable of keeping time with sufficient accuracy as to be useful for determining longitude.

case is lost, so H-1 sits, like H-2 and H-3, with its innards exposed for all to see, a Rube Goldbergish contraption of spindles, bars, knobs, and levers. It gleams of brass, but the main gears are wooden. The clock's face, with four dials, is elaborately engraved; the rest of the timepiece has a no-nonsense look about it, as if it were a miniature model for some fantasy factory cranking out interchangeable parts. And in a sense that is what it is, the products being interchangeable instants of time. The clock still runs. It is wound each morning by a member of the museum staff. Two oscillating pendulums, with brass balls at their ends and connected by coil springs, rock back and forth with a two-second beat. It is hard to imagine this thing in a captain's cabin on a ship at sea, the ship heaving and groaning in a storm, but such was H-1's fate. Harrison was ordered by the Admiralty to accompany the clock on a voyage to Lisbon and back. He suffered terribly from seasickness, but the clock kept almost perfect time.

Still, Harrison wanted better. He sought perfection more than he sought the twenty thousand pounds promised by the Board of Longitude. So here in their cases at the Royal Observatory are H-2 and H-3, each one somewhat more compact than its predecessor, each incorporating new inventions intended to make its pulse independent of grit, grime, heat, cold, moisture, dryness, tossing, turning, tightly wound spring, run-down spring. H-2 and H-3 beat in unison with H-1, all three clocks keeping the same two-second rhythm. There is something hypnotic about their goings; one soon finds oneself rocking or nodding in time with the clocks. H-2 and H-3 have no wooden gears; they are contrived completely of brass and steel. In retrospect one wonders how Harrison could ever have imagined that these exquisite but complex monsters could ever become standard issue in His Majesty's navy.

Then comes H-4. It is as if suddenly it dawned upon Harrison that he had for decades been chasing a folly: A small, high-frequency oscillator (five beats per second) might keep better time at sea than a big, clunky, brass contraption ever could. H-4 looks like a familiar pocket watch, although much larger in size, about six inches in diameter. Its works are hidden in an exquisitely engraved case, but inside are extraordinary innovations: jeweled pivots, a balance wheel, a bimetallic strip to compensate for temperature change, a miniature remontoire that rewinds eight times per minute to maintain a constant force from the driving spring. Unlike its companions, which tick and tock and spin and nod with not a little pomp, H-4 is inert. Although capable of running, it is so compact, so exquisitely put together with so many tiny parts, the museum authorities do not risk the abuse of cleaning and lubricating that would be required every few years if the clock were active.

The performance of this last artifact of Harrison's genius was *three times* better than the standard stipulated by the Longitude Act of 1714. It has been called "the most important timekeeper ever made," and it was not long before every ship went to sea with a direct descendant of H-4, the original marine chronometer, and every chronometer in His Majesty's navy kept Greenwich time. When Captain Robert Fitzroy set sail aboard HMS *Beagle* in 1831, to map the coasts of South America (and to bear young Charles Darwin to new realms of intellectual adventure), he carried with him *twenty-two* chronometers—some his own, some borrowed, some officially issued by the Admiralty—with which to ascertain the longitudes of distant shores.

Several times over the years I have visited Salisbury Cathedral, eighty miles to the southwest of London and not far from the great megalithic monument at Stonehenge. It is one of the most graceful of medieval cathedrals. As an added attraction, it is home to the oldest mechanical clock in Britain, probably the oldest mechanical clock in the world with most of its original parts and in working order. (It might reasonably be argued that Stonehenge, constructed thousands of years ago in prehistoric times, is the oldest "clock" in Britain; certainly the stones are aligned to mark the peregrinations of the Sun.) The Salisbury Cathedral clock was constructed in about the year 1386, of wrought iron, by an unknown craftsperson. It is about as big as a steamer trunk—an assembly of clunky gears in a boxy iron frame, driven by a falling weight and controlled by an escapement mechanism. The clock has no dial. It strikes the hour on a cathedral bell, as it faithfully did for nearly five hundred years until it was replaced in 1884. In 1956 it was repaired and set up on public display in the nave of the cathedral, where again it whirrs away like some goofy Brobdingnagian windup toy, the great-granddaddy of all subsequent mechanical timekeepers.

The Salisbury clock (there is another slightly later clock by the same craftsman at Wells Cathedral) served a religious rather than secular purpose—designating times for prayer—but it can also be thought of as a hinge of history, the physical embodiment of a new kind of time, each *tick* precisely like any other, gravity pulling inexorably on the driving weight, the vanes of the flywheel whirling first one way, then the other, mechanically, incessantly. This is the kind of time that gave a heartbeat to the European Renaissance, to the rise of a secular middle class, to entrepreneurship, to technical and scientific innovation, and ultimately to Newtonian time—*tick, tick, tick, tick*—abstract, mathematical, uniform, and eternal. Prior to the advent of mechanical clocks, time was conceived as the cyclical repetition of eternal archetypes: the annual and daily cycles of the Sun, the drama of fall and redemption. Within this older concept of time every person acquired his or her meaning only in relation to the archetypal narrative. Tradition and authority ruled, not by the exercise of overwhelming power but by divine right. The art and architecture of Salisbury Cathedral, like that of all churches of the European Middle Ages, was iconistic, symbolic, a coded characterization of the great chain of being that assigned every person (and thing) to a preordained place in the chain. But from 1386 onward, hidden away in a tower of the cathedral was this new thing, this wrought-iron box of spinning gears, ticking out a kind of time that has no beginning or end, that makes no reference to story, a great swimming continuum in which there are no borders, no constraints, and anything is possible.

There is a wonderful novel by Orhan Pamuk, Turkey's most popular contemporary writer, that takes us to that moment in world history when Europe crossed an intellectual divide from which there would be no turning back. Part murder mystery, part love story, part historical fiction, Pamuk's novel is called *My Name Is Red* and is set in Istanbul in the late 1590s. The Ottoman sultan Murat III has secretly commissioned a book that will celebrate his life and empire, to be illustrated by a group of master miniaturists, men trained in the artistic styles of the great traditional masters of Islamic text illumination. Why secrecy? The illustrations will be in the new European style of realistic representation, with shadow, perspective, and all the other tricks-in-trade of European Renaissance art—all heretical by Islamic standards. Shockingly, the book will also include a recognizable portrait of the sultan himself, not as a stylized appendage to Allah's word but as an object of admiration in itself. Portraiture, of course, had recently been brought to a high level of accomplishment in Europe; think, for example, of Hans Holbein's familiar portrait of Henry VIII. Within Sultan Murat's secret book, innovation confronts tradition, secularism confronts theocracy, individual artistic style confronts anonymous conformity to established modes of expression. Soon, two men are dead, and we have a baffling murder mystery on our hands that is not resolved until the final pages of the novel.

Pamuk's story concerns itself with art, but of course something else, not unrelated, was happening in Europe in the 1590s. Astronomers debated the truth of the Copernican system of the world, which removed the Earth (and humankind) from the center of the universe. Anatomists dissected the human body and used their careful observations to challenge ancient learning. Galileo began his studies of terrestrial motion. Soon the telescope and microscope would reveal new worlds, William Harvey would discover the circulation of the blood, and William Gilbert would explain the magnetic influence of the Earth. This upheaval in science can trace its beginning to art. Once an artist such as Albrecht Dürer could take as his subject a single rabbit or patch of weeds, and describe with lifelike realism every hair and whisker, every leaf and stem, the Scientific Revolution was inevitable. Once an artist, such as Dürer, prominently *signed* his work and took pride in his own *individual style,* the Reformation and collapse of monolithic theology was inevitable. With the Renaissance, Europe embraced progress, individual creativity, and empirical learning, and turned its back on tradition, religious conformity, and the authority of the past. Meanwhile, beating the rhythm for this transformation, were the Salisbury clock and its descendants, such as Tompion's and Harrison's timekeepers at the Greenwich observatory. European culture had broken free of the ancient myth of eternal return. Cyclic history, always turning upon itself, was finished. The future was henceforth wide open, and progress was in the wind.

As the sixteenth century began, Islamic civilization was experiencing a golden age, and one might reasonably have thought that the East was destined for cultural and military dominance over the West. It was not to be. The Turks were turned back from the walls of Vienna in 1529 and beaten at sea at Lepanto in 1571. But it was in the realm of ideas, not on the battlefield, that Europe gained its primary ascendancy. In his novel Pamuk describes a large mechanical clock with statuary sent as a gift to Sultan Murat III by England's Queen Elizabeth I, meant to represent, presumably, the best of European scientific, technical, and artistic innovation. Will Islam follow Europe's lead? Will Murat's illustrated book, in the European style of realistic representation, set a new standard of artistic illustration? Murat dies. His less forward-looking successor, Ahmet I, takes a mace to Elizabeth's gift clock and bashes it to pieces in the name of Allah—and returns Islamic book illustration to slavish imitation of the past. Pamuk's wonderfully original whodunit evokes a moment in Islamic history replete with all of the conflicted loyalties—to past and future—that are Islam today.

The ticking clocks in the museum of the Royal Observatory at Greenwich are more than a successful solution to the problem of longitude. They symbolize a kind of time that has been separated from human history once and for all—Newton's time, time without beginning or end, time that flows equably, mathematically, time for which any human clock can only be a pale imitation. Harrison's clocks gained or lost only a fraction of a second a day; present-day atomic clocks might gain or lose a second in 10 or 20 million years. For thousands of years the spinning Earth defined the day, hour, minute, and second. We now know that the Earth's rotation rate is erratic, even from year to year, and over the long haul is slowing down. Each century the length of an Earth day increases by a millisecond or two. Not much, but detectable with atomic clocks. Our new official international definition of the second—the duration of 9,192,631,770 periods of the radiation corresponding to the transition between the two hyperfine levels of the ground state of the cesium 133 atom at rest at zero degrees absolute temperature—makes no reference to the spinning Earth, and would presumably make sense to any sufficiently advanced civilization anywhere in the universe.

It is logical to believe that the eighteenth century's obsessive search for a solution to the problem of longitude led inevitably to James Hutton's concept of geologic time—cosmic history independent of human history—and finally to Charles Darwin's evolutionary story of life on Earth. Certainly, no part of human intellectual history is independent of any other. The whole point of Harrison's endeavor was to make a clock whose running was independent of anything the elements might hurl against it. Not only did his clocks keep track of a ship's position at sea; they symbolically provided an endlessly ticking metronome against which scientists of the nineteenth century might measure the rise and fall of mountains and the transmutation of species.

And something else. Harrison's H-4 coincided with the beginning of the factory system and the manufacture of interchangeable parts. It would be difficult to overestimate the importance of interchangeable parts in creating the modern world. Consider for a moment that until the late eighteenth century every part of every human artifact was made expressly for the object of which it was a part. The idea that a trigger mechanism, say, from one firearm might fit another firearm was inconceivable. Only when time and space were imagined to be made of interchangeable units (seconds, meters) was mass production possible. And the contribution of H-4 to the Industrial Revolution was not only conceptual. Once Harrison had shown the way, the demand for marine chronometers was so great that clockmakers could only meet it by farming out the making of standard parts. Chronometers were among the first artifacts manufactured with interchangeable components. The idea of interchangeability was in the air. When Thomas Jefferson wrote, "We hold these truths to be self-evident, that all men are created equal," he was not writing in an intellectual vacuum.

The marine chronometers of the British merchant and military fleets kept Greenwich Mean Time (GMT); that is, twenty-four clock hours correspond to the length of an *average* solar day. Because the Earth revolves about the Sun on a tilted axis and its orbit is not quite circular, the length of a solar day as measured by a sundial is not constant throughout the year. The time the Sun crosses a local meridian (sundial noon) can vary from a clock keeping average solar time by as much as sixteen minutes, a difference known as "the equation of time." John Flamsteed, the first astronomer royal, was the first to describe exactly the equation of time, and a mariner had to take it into account, according to the date, when comparing his chronometer against local Sun time.

Meanwhile, on land, every town and village set their local clocks against a sundial; that is, every town and village kept their own local solar time. When it was noon in London, a clock in Salisbury read eleven fifty-three and a clock in Penzance near the tip of Cornwall read eleven thirty-eight. And so on. This worked well enough until the railroads came, with their need for published schedules, and it became a practical necessity that all clocks in Britain keep the same time. And so the Greenwich observatory began distributing Greenwich Mean Time throughout the British Isles, by gunshots, by time balls, by physically carrying clocks from place to place, and finally by telegraph. By 1855 virtually every public clock in Britain displayed Greenwich time, and sundials became decorative objects only.

The situation was rather more complicated in the United States, where dozens of railroad companies each kept their own time standard. Britain is a small enough nation so that no community's GMT clock is more than a half hour or so different from sundial time, but if all communities in the United States kept the same time—Washington Mean Time, for example—clocks on the West Coast would be *hours* out of sync with the Sun, an unacceptable situation. To solve this problem a New York professor named Charles Dowd introduced the idea of time zones, each fifteen degrees (one hour mean solar time) wide, within which all clocks would keep mean solar time at the center of the zone. No one's clock within a time zone would be more than a half hour different from local solar time. Dowd's system was adopted in the 1880s, and—in an extraordinary show of internationalism—the United States and Canada opted for time zones anchored on Greenwich, England, rather than on Washington or Ottawa. Other nations soon followed. Christopher Wren's observatory on the hill in Greenwich Park took on a prominence that extended right around the world.

Key to establishing any system of longitude or time zones is a transit circle, a telescope mounted to swing in a precisely north-south plane, so that the passage of stars, Moon, or Sun across a local meridian can be measured. Wren's original observatory building at Greenwich was not practical for this purpose, and Flamsteed, the first astronomer royal, set up his transit instrument in a small shed in his garden. When Edmond Halley took over as astronomer royal, he noted that the brick wall upon which Flamsteed had mounted his instrument was beginning to subside, and he had a new meridian wall constructed slightly to the east. Subsequent improvements in transit instruments continued this pattern, and today one walks through a series of rooms at Greenwich containing the transit instruments of successive astronomer royals—Flamsteed, Halley, Bradley, and Airy—each a few feet east of the previous instrument and each of which for a time defined the British prime meridian. It is the last of these meridians, that of George Biddle Airy, that was recognized in 1884 as the prime meridian of the world. Today, the plane of Airy's transit instrument is prominently marked in the courtyard of the observatory, and visitors from all over the world have their photographs made standing with one foot to the east and one foot to the west of "Greenwich."

Somewhere among the boxes of photographs in my home is a picture of my own oldest three children standing in a row straddling the line. When I first visited the observatory in 1968–69 (the year of the photograph), it was a quiet place, not yet the bustling tourist attraction it is today. When I visited the observatory on my meridian walk, by midday the courtyard swarmed with people from (it seemed) every nation of world. It was deeply satisfying to see Asians, Africans, Australians, North and South Americans, and, of course, Europeans all agreeing upon one thing at least: the line across the courtyard by which everyone on the planet sets their maps and clocks, a triumph of scientific thinking over the deep-seated tendency to see ourselves at the center of the world, and a hopeful sign that someday those differences of religion, politics, and race that still divide us so violently might fade as blessedly into inconsequence as did our previous squabbles over whose maps and clocks would properly designate space and time.

6

COSMIC SPACE

If the Royal Observatory at Greenwich has a royal star, it has to be Gamma Draconis, also known as Eltanin, "the dragon's head." Eltanin is not an especially bright star, in a constellation, Draco, the Dragon, without any bright stars. But it does have a claim to local fame: It passes directly overhead London once each day. Let's say you live in the seventeenth century and want to measure the position of a star with exceeding accuracy. There are advantages to choosing a star that passes near the zenith. First, gravity defines the zenith exactly; you need only align a telescope with a plumb bob to be sure that it points vertically. Second, the star's light passes perpendicularly through the Earth's atmosphere, so it will not be subject to refraction, the bending of light that occurs when light passes obliquely through any transparent medium. In 1669 Robert Hooke, Newton's nemesis, proposed to observe Gamma Draconis throughout the course of a year, measuring its precise position with respect to the zenith. What he hoped to discover, he knew, would make him famous: unambiguous proof that the Earth moves through space as Copernicus had proposed. Furthermore, if successful, he would be the first to measure the distance to a star.

Not that Hooke needed another feather in his cap. He was already acknowledged as one of the cleverest natural philosophers of his time, a new Leonardo. In 1662 he had been appointed curator of experiments for the newly created Royal Society, and in the following years he contrived an extraordinary range of experiments and inventions. Vacuum pump, air compressor, universal joint, iris diaphragm (today used in cameras), spiral spring for clocks, barometer, hygrometer, wind gauge, clock-driven telescope—his fertile mind cranked out one useful device after another. His international reputation was assured with the publication in 1665 of *Micrographia,* a wondrous compendium of the world of the very small revealed by the newly invented microscope, with exquisite drawings. The diarist Samuel Pepys called it "the most ingenious book that I have ever read in my life."

Strange fellow, Hooke. A sickly, stooped, pop-eyed man, afflicted with an assortment of uncertain maladies, including, no doubt, hypochondria. A nervous, cantankerous insomniac, jealous of his fame, quick to put down others. He seems to have intuited the inverse square law of gravity before Newton, but he lacked the patience and mathematical skill to do what Newton did: apply the law to both terrestrial and celestial motions. Today his name appears in science books only in association with the elastic law of springs, not a particularly fundamental insight into the workings of nature. As with Leonardo, his restless mind flitted so nimbly from topic to topic that he seems never to have followed through any idea fully. And so it was with his 1669 proposal: to measure stellar parallax.

Parallax is an *apparent* change in the position of an object when viewed from two different places, a fancy word for a common notion. Hold your finger up in front of your nose. Now look at it first with one eye, then the other. Notice how your finger seems to move against the more distant background. Notice also that as you move your finger farther away from your nose, the apparent shift becomes less. *The greater the distance of the observed object, the less the parallax.* For example, figure 6-1 illustrates the Moon viewed from London and Paris against the background of the more distant stars. If the distance from London to Paris is known (the baseline, 210 miles) and the apparent angular shift of the Moon is measured (the parallax angle), then it is a simple matter to calculate the distance to the Moon. But note that figure 6-1 is not to scale. The actual Moon and stars are much farther away compared to the London-Paris distance. Figure 6-2 shows the Moon against the background of stars as if it were actually observed *at the same time* at London and Paris. The Moon in each "photograph" is almost exactly one half a degree wide. If you look carefully, you will see that the Moon is in a slightly different position relative to the background stars, not because the Moon has moved but because we are viewing it from different places. The apparent shift is a small but measurable fraction of a degree—the parallax angle.

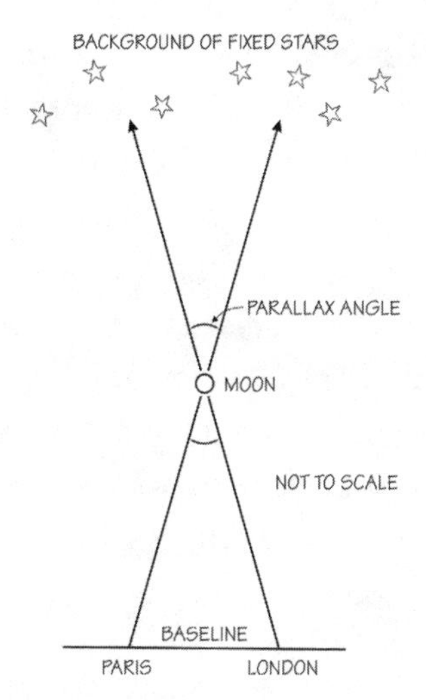

FIGURE 6-1. Schematic drawing showing the parallax of the Moon viewed from London and Paris.

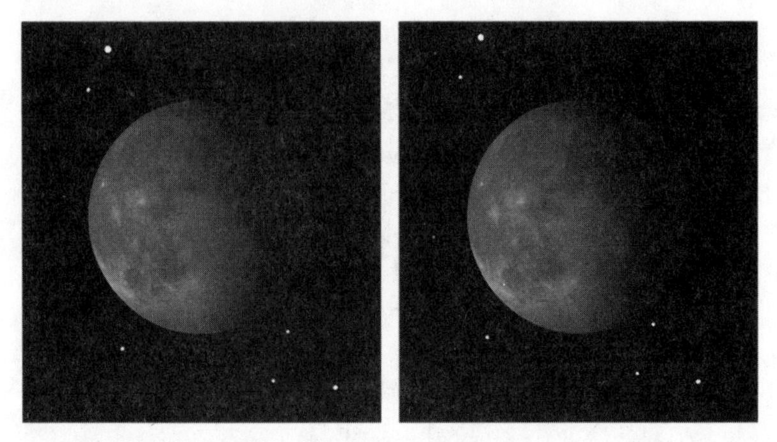

FIGURE 6-2. The Moon observed at the same time from London and Paris. Note the small shift against the background stars—an example of parallax.

But a larger baseline than the distance from London and Paris is necessary to have any hope of measuring the parallax of stars. When Aristarchus, and later Copernicus, suggested that the Earth moves in a great orbit around the stationary Sun, it became at least *theoretically* possible to measure the distance to the stars. When viewed from opposite sides of the Earth's orbit, the stars should show a shift in apparent position with respect, say, to the zenith. The relevant diagram is the same as that of figure 6-1, except instead of the base of the triangle being the distance from London to Paris it is now *the diameter of the Earth's orbit about the Sun,* 186 million miles! The Alexandrian astronomers were aware that the stars showed no perceptible annual parallax with respect to each other or to any other reference point in the sky, or at least none that they could measure, which is surely one reason they rejected Aristarchus's notion of a moving Earth.

The other possibility is that the stars are exceedingly far away *compared even to the Earth's orbit,* an idea that seems to have frightened everyone in the ancient world except Aristarchus. The absence of perceptible stellar parallax means that if you accept the idea of a heliocentric universe, you must also buy into a universe that is staggeringly large—a universe in which the gods become unpleasantly remote from the arena of human affairs. Thus the charge of impiety against Aristarchus. Copernicus too knew he was treading on dangerous ground. Giordano Bruno went to the stake exulting in the infinity of space. Galileo was brought to his knees before the assembled officers of the Inquisition.

Of course, by the time of Newton and Hooke, most scientists had become convinced that Copernicus was right, and had grown used to the idea that the stars were very, very far away. But how far? There were two ways to find out. One way was to assume (without proof) that the stars are other suns, and calculate how far away they must be to appear as bright as they do in the sky. (The intensity of a point source of light diminishes with the square of the distance.) Isaac Newton, Christian Huygens, and others pursued this method and estimated that the star Sirius, the brightest star in Earth's sky and therefore possibly the closest, was hundreds of thousands times farther away from the Earth than the Earth is distant from the Sun! To give you an idea of scale, this means that if the Earth's orbit were the size of a compact disc (CD), the nearest star would be twenty miles away! And so did the cosmic egg of the ancients, and of Dante's *Divine Comedy,* yield to a yawning, silent abyss.

It turns out, of course, that the stars *are* other suns, as Newton and Huygens supposed, but the assumption that all stars have the same intrinsic brightness is wrong, as was the assumption that the brightest star is also the closest. We now understand that some stars are vastly more luminous than the Sun, and some are less luminous. Nevertheless, the method of gauging stellar distance by apparent brightness did give *very roughly* the correct distance to the nearest stars.

The second method of measuring stellar distances—parallax—makes no assumptions about the intrinsic brightness of stars. The method is direct and unambiguous: measure the parallax angle and calculate the long side of the triangle. Unfortunately, given the technology of Newton's time, measuring such a tiny angular shift in the apparent position of a star was impossible, even as it was at the time of the Alexandrians. (Imagine looking at an object twenty miles away first with one eye, then the other.) But this is exactly what Robert Hooke proposed to do: to see Gamma Draconis make an apparent shift, back and forth, as the Earth went around the Sun.

He cut holes through the interior ceilings and roof of his apartment at Gresham College in London, and set up a thirty-six-foot-long vertical telescope, aligned with a plumb bob. With this he watched Gamma Draconis transit near the zenith (the point directly overhead), carefully ascertaining its angular distance from the zenith point. He made measurements on July 6, July 9, August 6, and October 21, all in the year 1669, and *thought* he observed a tiny shift of a hundredth of a degree. But he was not at all certain that his apparatus worked reliably, and had little confidence in this result. After less than a handful of observations, he abandoned the project due to "inconvenient weather and great indisposition to my health." And just as well. It turns out that Gamma Draconis's parallax is less than a thousandth of what Hooke thought he had measured.

Still, give Hooke his due. He had a go at measuring the distance to a star and established an experimental plan that others would follow. If he didn't find the distance to Gamma Draconis, or see the annual wobble in the star's apparent position that proved the Earth moves, he at least confirmed more reliably than anyone before that *if the Earth did move,* the stars were very far away indeed.

More than a century would pass before the German astronomer Friedrich Bessel finally succeeded in measuring the parallax of a star, with a purpose-built telescope vastly improved over the instrument used by Hooke. The star whose parallax Bessel measured in 1838 was not Gamma Draconis, nor was it a star near his German zenith. The star was an altogether more nondescript candidate, named 61 Cygni. He chose this star not because it was bright but because it had a relatively large so-called proper motion; that is, an actual, rather than apparent, motion through space. Bessel guessed that stars with large proper motions when viewed from Earth will be likely to be among the nearest stars, and in the case of 61 Cygni he was right. It is our eleventh nearest neighbor (counting multiple star systems as one) of the hundreds of billions of stars in the Milky Way Galaxy.

We now know that Hooke could never have succeeded with Gamma Draconis, because the star is too distant—150 light-years away (900 trillion miles), more than ten times farther away than 61 Cygni—for Hooke to have seen a parallax shift. If the Earth's orbit were the rim of that compact disc I mentioned before, with the Sun a tiny dust mote at its center, then 61 Cygni would be another dust mote *twenty-four miles away.* The apparent angular shift Bessel succeeded in measuring is that subtended at the eye by the radius of a compact disc twenty-four miles away—about one ten thousandth of a degree—an achievement that pushed the limits of astronomical precision.

By the end of the nineteenth century, sixty-two years after Bessel's triumph, the parallax of fewer than one hundred stars had been measured. Today, with purpose-built satellite telescopes, the distances to more than a *billion* stars have been directly determined by parallax—including, of course, Gamma Draconis and 61 Cygni—but these are still only our nearest neighbors in the Milky Way Galaxy.

Although Robert Hooke failed to measure stellar parallax, I doubt very much if he would have been surprised at the scale of the universe as it has been revealed today. He gave up quickly on Gamma Draconis—which turned out to be a smart move—and busied himself with other schemes and diversions. When Charles II decided to establish his Royal Observatory, in 1675, Hooke was there to have a hand with Christopher Wren in its design and construction. And before we leave the Royal Observatory to continue our walk along the meridian, there is one more person we should meet: the third astronomer royal, the Reverend James Bradley.

It would be hard to find anyone more *unlike* Hooke than Bradley. Bradley was as placid and steady as Hooke was cantankerous and flighty. Once Bradley got his teeth into a problem, he stuck with it until he found a solution. And the problem he took between his teeth was the very one that had defeated Hooke: the measurement of stellar parallax.

Moreover, Bradley thought Hooke was correct to choose a star that passes near the local zenith. He too set his sights on Gamma Draconis, but he was determined to build an instrument as rock steady and reliable as his own temperament. This he did and began a long series of observations of the star, checking and rechecking his instrument every step of the way. And indeed Gamma Draconis measurably moved, in an exactly *annual* cycle that confirmed that the apparent motion was an artifact of the Earth's orbit. But it was not a motion that could be explained by parallax; that is, it did not correspond to what would be expected if the star were viewed from different places in the Earth's orbit (the blinking eyes in my analogy). Bradley was baffled. He had no clue as to what was going on.

Then one day while sailing on the Thames, he happened to be watching the little wind vane flying at the top of the boat's mast. The vane, of course, was meant to tell the helmsman the direction of the wind. However, each time the boat put about, the vane shifted slightly, *as if the direction of the wind had changed.* But of course it was unlikely that the wind should change direction exactly as the helmsman decided to turn the boat. On questioning the sailors, Bradley was told that the vane indicated the combined velocities of the wind and the boat; that is, the boat's motion through the water was itself a kind of wind that affected the direction of the vane. *Aha!* Suddenly it dawned on Bradley what he was observing with Gamma Draconis: a combination of the Earth's velocity in its orbit and *the velocity of light.*

Consider this analogy. You are standing still in rain that is falling vertically. You are holding a long tube. The raindrops fall straight through the tube. Now start walking. If the raindrops are to fall through the tube, you will have to tip the top of the tube slightly forward; otherwise the drops will encounter the side of the moving tube while falling through it. It's the same reason you tip your umbrella forward when you walk in rain. And so it is with starlight falling vertically on Bradley's telescope. As the Earth moves in its orbit, first one way, then the other, the telescope has to be tipped to bring the light to a focus, which makes the position of the star *seem* to change. What Bradley discovered was not parallax but what is called the *aberration of starlight.* He had not measured the distance to a star, but he had found irrefutable evidence of the Earth's motion. The last recalcitrant anti-Copernicans were persuaded. It was no longer possible to believe that the Earth rested immobile at the center of the cosmos.

Yet Bradley persisted in his observations of the star, seeking the wobble *on top of the wobble* that would be parallax—and found still another complication. The Earth's axis wobbles ever so slightly due to the Moon's gravitational tug on the Earth's slightly oblate form, an effect called *nutation.* This too makes a star's position seem to change. Gamma Draconis does a rather complicated dance in the sky, but try as he might, Bradley found no part of its jiggly behavior that corresponded to parallax. He was confident that his telescope could have detected a change in position as small as one arc second (1/3600 degree), which implies that Gamma Draconis had to be at least four hundred thousand times farther away from Earth than Earth is distant from the Sun. But how far Bradley could not say. He did conclude that choosing a star for parallax measurement just because it happened to lie near the zenith was foolish. There is no reason to assume in advance that Gamma Draconis, for instance, is a near neighbor. It would be better, he guessed, to pick a bright star such as Sirius, on the assumption that bright means near. But then, away from the zenith, the measurement of tiny angles is complicated by refraction in the Earth's atmosphere.

Galileo had proposed a way of getting at parallax that Bradley now embraced as perhaps offering the best promise for astronomers. First, search for two stars lying very close together in the sky along the same line of sight—with one member of the pair much less bright than the other. The less bright star presumably lies much farther away than the brighter member of the pair. Measure the change in position of the nearby star with respect to the more distant star. All of the problems that frustrate the search for parallax—atmospheric refraction, uncertainties of telescope alignment, aberration, nutation, and so on—would presumably affect both stars equally. Only parallax, which depends on distance, would change the positions of the two stars *relative to each other;* that is, the nearby star would show a larger annual shift in position than the more distant star. So the faraway star provides a convenient reference point for measuring the parallax wobble of the closer star. Or so Bradley was convinced as his own long and fruitful attention to the jiggle of stars came to an end.

William Herschel (1738–1822) took up the challenge. Herschel was a German musician who came to England at age nineteen and quickly established himself as the preeminent astronomer in the land, building telescopes that outperformed the instruments at Greenwich. He settled in Slough, west of London, and eventually constructed there an instrument with a mirror four feet in diameter mounted in a cast-iron tube forty feet long, which for half a century would be the largest telescope in the world. King George and the archbishop of Canterbury paid a visit; Herschel's behemoth was something of a tourist attraction. With his excellent instruments and the assistance of his capable sister Caroline, Herschel amassed an array of discoveries that no single astronomer has equaled before or since, including a new planet, Uranus, and thousands of nebulae and star clusters. And double stars. Herschel cataloged the locations of a thousand double stars, and following Bradley's suggestion he undertook to observe over time the relative positions of doubles with widely divergent brightnesses. He did not find parallax but discovered something almost as interesting. It turned out that many of his double stars were *true binary systems,* two stars bound together in a gravitational dance, and not merely stars that happened to lie along the same line of sight. Since these pairs were bound together, yet of widely different brightness, Herschel proved once and for all that all stars are not equally bright. Apparent brightness is not a reliable indicator of distance.

And so stellar parallax remained devilishly elusive, but the search for parallax led astronomers on a merry chase into a universe of surprising variety and dimension. The distances to stars weren't measured directly until Bessel's success in 1838, but by then William Herschel (his son John also became a famous astronomer) had revealed the shape and size of the Milky Way Galaxy, and cataloged fuzzy spots in the sky that would later prove to be other galaxies. By the time William Herschel died in 1822, it had become manifestly clear that the visionary philosopher-poet Giordano Bruno was right: The Sun is a typical star in a universe of uncountable stars. The human abode is a dust mote in a cathedral swirling with dust motes. No choirs of angels look down on our human travails from Dante's heaven just up there beyond the dome of stars. The heavens reach on endlessly, vast and empty, sprinkled here and there with other suns, other planets, other galaxies.

There have been fifteen astronomers royal since the founding of the Royal Observatory at Greenwich. It is a post that still exists, and as I write, the present occupant is Sir Martin Rees. But the observatory on the hill in Greenwich Park plays an increasingly irrelevant role in British astronomy. By the middle of the twentieth century light and air pollution had become significant problems, as the city of London reached out to embrace what had previously been a sleepy suburb. In 1946 the observatory moved to a new home at Herstmonceux Castle in Sussex, not so far from the prime meridian. The astronomer royal left Greenwich in 1948, and the last astronomical observations at Greenwich were made in 1954. The buildings on the hill are now part of the National Maritime Museum and stand as a proud monument to humankind's search for our place in space and time.

In 1990 the Royal Observatory moved again, this time to Cambridge, but only as a base of operations. Inclement British weather, plus light and air pollution, means that British optical astronomers must now go far afield to do important research, most commonly to the Northern Hemisphere Observatory in the Canary Islands. On the other hand, clouds, ambient light, and air pollution do not significantly affect *radio* astronomy. For a radio astronomer, day and night, cloud or clear, are all the same; long-wavelength radio waves pass as easily through clouds as they do through the walls of your house. With so many clouds to contend with, it was perhaps inevitable that the British would be among the pioneers of radio astronomy. The twelfth astronomer royal, Martin Ryle, led the way during the years immediately after the Second World War. As in so much of British science, the Cavendish Laboratory at the University of Cambridge figured prominently in the building of the world's first radio telescopes.

So I had one last destination on my meridian walk. From Greenwich a pedestrian tunnel took me under the Thames. Then I followed a long-distance footpath north along the River Lee, which lies almost exactly along the line of zero longitude. At Ware the river jags westward with its bankside footpath, so I made my way through the rolling landscape of Hertfordshire and Cambridgeshire. The walking was by no means unpleasant, the public pathways common and convenient. As I approached the outskirts of Cambridge, hard by the meridian line, I could see on the horizon the multiple white dishes of the Mullard Radio Astronomy Observatory of the University of Cambridge, cocked like so many ears to the heavens. The astronomers who tend these instruments are investigating star births and deaths, galactic evolution, the big bang, quasars, pulsars, and black holes.

Celestial objects emit radiation in all parts of the electromagnetic spectrum, from long-wavelength radio waves to very short-wavelength gamma rays. Only the visible part of the spectrum (the colors of the rainbow) and radio waves penetrate the Earth's atmosphere, so only those parts of the spectrum are available to ground-based astronomers. But all parts of the electromagnetic spectrum carry useful information, and so today gamma-ray, x-ray, ultraviolet, infrared, and microwave telescopes circle the Earth above the atmosphere, gathering clues to the nature of the universe. Physicists also predict the existence of gravitational waves, ripples in the fabric of space-time caused by catastrophic events across the galaxy or deep in time, and although these have not yet been detected, the search is on.

No space telescope has provided more stunning views of the cosmos than the Hubble Space Telescope. As I write, NASA has just released a new Hubble photograph called the Ultra Deep Field. To make the photograph, the telescope was focused on a tiny part of the sky that could be covered by the intersection of crossed straight pins held at arm's length. The exposure was one million seconds (278 hours), requiring more than four hundred orbits of the telescope. This is the deepest view of space ever achieved. More than ten thousand galaxies are visible in the photograph, the most distant more than 13 billion light-years away. The light from these most distant galaxies began its journey when the universe was only 5 percent of its present age.

Try this: Go out tonight and hold up crossed straight pins at arm's length against an utterly dark part of the sky. Now try to imagine the ten thousand galaxies the Hubble sees within that tiny square of darkness, galaxies reaching back to the dawn of time. Each of those galaxies contains tens or hundreds of billions of stars (it is the brightest galaxies we are seeing), and each of those stars may have planets. Not even Giordano Bruno could have imagined a universe such as the one we find ourselves a part of.

How old is the universe? The amazing thing is not the answer—approximately 13.7 billion years—but the fact that there *is* an answer.

Early in the past century, the most widely held view among scientists was that the universe is eternal, with no beginning or end—the so-called steady-state universe. Then, in the 1920s, astronomers working at the new Mount Wilson Observatory in California made an astonishing discovery: The universe is expanding. The galaxies are racing away from each other. And if the galaxies are moving apart, they must have been closer together in the past. Theoretically, we can run the movie backward using the laws of physics to tell us what happens. The galaxies converge. The density of matter increases, and the temperature soars. Atoms dissolve into their constituent parts. Mass becomes pure energy. Run the movie 13.7 billion years or so into the past, and the whole thing—the entire universe of galaxies we observe today—collapses into an infinitely small, infinitely dense, infinitely hot mathematical point. Time goes to zero. The universe begins!

Many astronomers of the 1920s were not happy with what the data was telling them. An eternal universe is hard to imagine, but a universe that has a beginning is even harder to imagine. Where did it come from? What caused it to begin? It is so much easier to assume that the universe has existed forever.

But the data from Mount Wilson was not to be denied. The speed of the galaxies away from us can be measured from a stretching of their light (the same principle a policeman uses to check your car's speed with a radar gun). The distances of the galaxies are estimated from the apparent brightnesses of stars within the galaxies—supernovas or certain kinds of variable stars whose absolute brightness is known—or from the apparent brightnesses of entire galaxies: the less bright (on average), the farther away. Put it all together, and we are led inevitably to the big bang.

Until recently, the exact age of the universe has been beyond our grasp, mainly because of uncertainties in the distances of the galaxies. Estimates of the universe's age varied by several billions of years. Fortunately, the age of the universe suggested by the receding galaxies is satisfyingly greater than the age of the Earth—4.6 billion years—which is known with more accuracy. If it had turned out the other way around—a universe younger than the Earth—we would know something was terribly wrong with the science.

Since the 1920s, several ways of estimating the universe's age have been devised. One of them uses the cooling rate of white dwarf stars, the slowly fading embers of stars that are no longer producing energy. Another method relies upon the measured abundances of radioactive thorium in the atmospheres of ancient stars (a variation of the carbon-14 "clock" used by archaeologists on Earth). All of the methods are to some extent inexact, but all converge on an age for the universe somewhere between 10 and 16 billion years.

Not long ago, an international team of astronomers working at the European Southern Observatory in Chile discovered the signature of two different radioactive elements, thorium and uranium, in the spectrum of a star called CS31082-001. The presence of two "clocks," started at the same time and running at different rates, sharpens the estimate of when the radioactive elements were created, presumably in supernova explosions early in the universe's history. The age of the radioactive elements, according to this new study, is 12.5 billion years, plus or minus 3.3 billion years—agreeably consistent with earlier estimates.

More recently comes the best guess yet for the age of the universe. The Wilkinson Microwave Anisotropy Probe (WMAP) is a purpose-built satellite telescope designed to produce an image of the infant universe when it was less than four hundred thousand years old—a time before stars and galaxies were born, when the universe was a blazing hot cauldron, a plasma, of energy and incipient matter. The big bang set the plasma ringing like a bell, and WMAP studies the "tones." Just as the shape of a bell determines its tone, so do variations in microwave radiation reveal the character of the early universe. And its age. With WMAP data, astronomers can pin down the rate of the universe's expansion and confidently say that the cosmos is 13.7 billion years old, plus or minus a few hundred thousand years.

A human lifetime is almost unimaginably brief compared to the age of the universe. Imagine a human lifetime to be represented by the thickness of a single playing card. Then the age of the universe is a pile of cards forty miles high, or roughly the distance from London to Cambridge. Hold a playing card between your thumb and forefinger and consider the long walk between the two cities, and you will begin to understand the difference between human time and cosmic time. As a species we can take considerable pride that within a human lifetime we have garnered a robust picture of how the universe began, and made an increasingly satisfying guess for when it happened.

Think of it this way. A lucky mayfly might live for an hour. By analogy with a human lifetime and the age of the universe, it is as if mayflies, in their brief fling of a summer's evening, were able to figure out what was happening on Earth twenty-five thousand years ago!

The track of the prime meridian across England from Peacehaven in the south to the mouth of the River Humber in the north is nearly two hundred miles. If that distance is taken to represent the 13.7-billion-year history of the universe, as we understand it today, then all of recorded human history is less than a single step. The entire story I have told in this book, from the Alexandrian astronomers and geographers to the present-day astronomers who launch telescopes into space, would fit neatly into a single footprint. If the two hundred miles of the meridian track are taken to represent the distance to the most distant objects we observe with our telescopes, a couple of steps would take us across the Milky Way Galaxy. A mote of dust from my shoe is large enough to contain not only our own solar system but many neighboring stars.

We have come a long, long exhilarating way from the cosmic egg of the ancients.

EPILOGUE

The human journey to cosmic space and time was bedeviled at every step of the way by what the biologist Richard Dawkins calls the Argument from Personal Incredulity: If it seems impossible to believe, it must be wrong. Aristarchus confronted the incredulity of his contemporaries, as did Copernicus, Bruno, Galileo, and Darwin. The universe invariably turned out to be bigger and older than we had previously thought possible. The light-years and the eons are a rebuke to the limitations of our imaginations—and a tribute to the power of the boldest, most daring human thinkers to transcend "common sense."

We are the inheritors of this proud tradition. We stand under the dark night sky and let our imaginations follow the pointing shadow of the Earth into the inky depths—moon, planets, stars, galaxies, even the radiant energy of the big bang recorded by microwave satellites—through vertiginous empty spaces toward the singular instant of creation. Into the star-flecked darkness we let our imaginations soar—paces, miles, thousands of miles, millions of miles, light-years, millions of light-years, billions of light-years—following an Adrienne's thread of theory, observation, and unquenchable curiosity invented 2,300 years ago in a glistening white city at the mouth of the Nile.

Let us end our trek along the Greenwich meridian at

Westminster Abbey, in London, only five miles from Greenwich, a short and popular boat ride along the Thames. The fourteenth-century abbey rises on graceful flutes and columns more than hundred feet, making it the tallest Gothic structure in the British Isles. The fan vaulting in the sixteenth-century Lady Chapel is a thing of almost miraculous beauty and soaring delicacy. The point of Gothic architecture, here and elsewhere, was to direct the worshiper's attention upward to a realm of heavenly splendor, away from the dismal squalor of the Earth. Life in medieval Europe was dangerous and grim. Even a slight knowledge of those violent, disease-ridden times makes clear why the silent, luminous, heaven-piercing spaces of Westminister Abbey were welcomed by medieval Londoners as a promise of something better.

But for all of its architectural grandeur, Westminster Abbey can be something of a disappointment for the twenty-first-century visitor. To an extent unparalleled in any other medieval cathedral I have visited, the abbey has been allowed to become a gaudy monument to posthumous vanity—to the centrality of self. The place is chock-a-block with outsized memorials and sarcophagi celebrating the lives and accomplishments of English men and women; sometimes it seems that the lesser the fame the more assertive is the monument. The effect of this worldly clutter is to make it almost impossible to appreciate the *upward and outward* aspirations the architecture was meant to evoke. This may be one reason why so many tourists are drawn to the Poets' Corner of the abbey, where Chaucer, Shakespeare, Dickens, and other literary lights are buried or memorialized. The memorials of the authors are relatively modest, as they should be; after all, art is its own memorial.

Some visitors to the abbey find their way to what might be called the Scientists' Corner, at a side of the nave near a corner of the quire. Here Isaac Newton is interred within a sarcophagus as gaudy as any other in the abbey, including a sculpted likeness of the great man himself looking pompously foolish in a Roman toga. The lengthy Latin inscription on his tomb begins, "Here lies Isaac Newton, Knight, who by a strength of mind almost divine, and mathematical principles peculiarly his own, explored the course and figures of the planets, the paths of comets, the tides of the sea, the dissimilarities in rays of light, and, what no other scholar has previously imagined, the properties of the colors thus produced."

The tombs and memorials of other scientists are more appropriately reticent. And what an assembly! Among them Charles Lyell, the father of geology, who inspired so many of the explorers we have met along our walk. William and John Herschel, who plumbed the depth of space. The physicists James Prescott Joule and George Stokes. Joseph Lister, the pioneer of antiseptic surgery. And of course, the greatest of them all, who lies beneath a dignified black slab inscribed with these few words: "Charles Robert Darwin. Born 12 February 1809. Died 19 April 1882."

Poor Darwin. He would perhaps be abashed to find himself in Westminster Abbey at all, so reclusive and retiring was he in life. And his doubts about traditional theology provide another incongruity to his final repose in this most eminent symbol of Anglican orthodoxy. But, in a sense, Darwin carried on the work of the architects and master craftsmen who built the Gothic churches.

"There is a grandeur in this view of life," Darwin wrote of evolution. In knitting the history of life, including our own species, into the space and time of the geologists and astronomers, Darwin helped to accomplish what the medieval builders sought in their own way to do: to lift our eyes from the confining circle of our birth and draw our attention to the light and glory of the cosmos.

ACKNOWLEDGMENTS

I am grateful to the many scholars and writers who did the academic legwork for my meridian walk. Many parts of my story have been told in greater detail elsewhere, and in "Further Reading" I have listed some books that may be of interest to readers of *Walking Zero.* My account of Thomas Huxley's talk at Norwich is essentially the same as the story I told in *Natural Prayers,* now out of print; the geography of my walk made it useful to retell the story here. As with my four previous books, I am greatly indebted to my wonderful editor at Walker & Company, Jacqueline Johnson. She is every author's dream editor. Thanks, too, to George Gibson, my publisher at Walker, for his continued confidence in my work, and the other talented people at Walker, especially Vicki Haire and Greg Villepique, who helped make *Walking Zero* a book. My son Dan Raymo of Platypus Multimedia prepared the illustrations. Thanks to my friend Barbara Estrin, for making available her flat in London as a base for my trek. My spouse, Maureen, read several drafts of the manuscript and offered her usual perspicuous comments; she wasn't with me as I tramped the meridian, but we have walked the long, long walk together.

ART CREDITS

Front map, figures 1-1, 1-2, 2-1, 2-2, 2-3, 2-4, 3-3, 3-5, 6-1, and 6-2 illustrated by Dan Raymo. The following images were illustrated by Dan Raymo, based on source material: Figures 2-5, 2-6, 2-7, and 2-8, adapted from original drawings by Chet Raymo. Figure 1-3, adapted from William Smith's *Atlas of Ancient Geography,* 1874. Figure 1-4, adapted from Girolamo Ruscelli's *La geografia di Clavido Tolomeo,* 1561.

Figures 3-1 and 3-2, photographs by Chet Raymo, used with the kind permission of the Natural History Museum, London.
Figure 3-4, reprinted from Gideon Mantell's *Geology of Sussex,* 1827.
Figure 3-6, courtesy of the British Geological Survey.
Figure 4-1, photograph by Chet Raymo.
Figure 4-2 and 4-3, reprinted from H. G. Wells's *Outline of History,* 1920.
Figures 5-1 and 5-2, photographs by Chet Raymo, used with the kind permission of the National Maritime Museum, Greenwich.
Figure 5-3, courtesy of the National Maritime Museum, Greenwich.

FURTHER READING

Prologue

Eliade, Mircea. *Cosmos and History; The Myth of the Eternal Return.* Translated by Willard R. Trask. New York: Harper, 1959.

Chapter 1. Mapping the Earth

Alder, Ken. *The Measure of All Things: The Seven-Year Odyssey and Hidden Error That Transformed the World.* New York: Free Press, 2002. The story of Jean-Baptiste-Joseph Delambre, Pierre-François-André Mechain, and the measurement of the French meridian.

Blaise, Clark. *Time Lord: Sir Sandford Fleming and the Creation of Standard Time.* New York: Pantheon, 2000.

Brown, Lloyd. *The Story of Maps.* New York: Dover, 1979.

Piaget, Jean. *The Child's Conception of the World.* Totowa, N.J.: Littlefield, Adams, 1969.

Chapter 2. The Earth in Space

Ferguson, Kitty. *Tycho and Kepler: The Unlikely Partnership That Forever Changed Our Understanding of the Heavens.* New York: Walker, 2002.

Galilei, Galileo. *Sidereus Nuncius, or The Sideral Messenger.* Translated by Albert Van Helden. Chicago: University of Chicago Press, 1989.

Heath, Thomas. *Aristarchus of Samos: The Ancient Copernicus.* Oxford: Clarendon Press, 1913.

Chapter 3. Antiquity of the Earth

McGowan, Christopher. *The Dragon Seekers: How an Extraordinary Circle of Fossilists Discovered the Dinosaurs and Paved the Way for Darwin.* Cambridge, Mass: Perseus, 2002.

McPhee, John. *Annals of the Former World.* New York: Farrar, Straus and Giroux, 2000.

Whitrow, G. J. *Time in History: Views of Time from Prehistory to the Present Day.* Oxford: Oxford University Press, 1989.

Winchester, Simon. *The Map That Changed the World: William Smith and the Birth of Modern Geography.* New York: HarperCollins, 2001.

Chapter 4. Antiquity of Man

Desmond, Adrian, and James Moore. *Darwin: The Life of a Tormented Evolutionist.* New York: Warner, 1991.

Edgar, Blake, and Donald Johanson. *From Lucy to Language.* New York: Simon and Schuster, 1996.

Huxley, Thomas. *On a Piece of Chalk.* Edited and with an introduction by Loren Eiseley. New York: Charles Scribner's Sons, 1967.

Spencer, Frank. *Piltdown: A Scientific Forgery.* Oxford: Oxford University Press, 1990.

Tattersall, Ian. *The Last Neanderthal: The Rise, Success, and Mysterious Extinction of Our Closest Human Relatives.* Boulder, Colo.: Westview Press, 1999.

Chapter 5. Cosmic Time

Gleick, James. *Isaac Newton.* New York: Vintage, 2004.

Pamuk, Orhan. *My Name Is Red.* New York: Knopf, 2001.

Pepys, Samuel. *The Shorter Pepys.* Edited by Robert Latham. Berkeley: University of California Press, 1985.

Sobel, Dava. *Longitude: The True Story of the Lone Genius Who Solved the Greatest Scientific Problem of His Time.* New York: Walker, 1995.

Chapter 6. Cosmic Space

Ferguson, Kitty. *Measuring the Universe: Our Historic Quest to Chart the Horizons of Space and Time.* New York: Walker, 1999.

Hirshfield, Alan W. *Parallax: The Race to Measure the Cosmos.* New York: W. H. Freeman, 2001.

INDEX